CHANGING HOW WE CHOOSE

CHANGING HOW WE CHOOSE

The New Science of Morality

A. DAVID REDISH

The MIT Press
Cambridge, Massachusetts
London, England

The MIT Press would like to thank the anonymous peer reviewers who provided comments on drafts of this book. The generous work of academic experts is essential for establishing the authority and quality of our publications. We acknowledge with gratitude the contributions of these otherwise uncredited readers.

This book was set in Adobe Garamond and Berthold Akzidenz Grotesk by Jen Jackowitz. Printed and bound in the United States of America.

Library of Congress Cataloging-in-Publication Data

Names: Redish, A. David, author.
Title: Changing how we choose : the new science of morality / A. David Redish.
Description: Cambridge, Massachusetts : The MIT Press, [2022] | Includes bibliographical
    references and index.
Identifiers: LCCN 2021060558 (print) | LCCN 2021060559 (ebook) |
    ISBN 9780262047364 (hardcover) | ISBN 9780262371438 (epub) |
    ISBN 9780262371445 (pdf)
Subjects: LCSH: Decision making—Moral and ethical aspects. | Social ethics.
Classification: LCC BF448 .R39 2022 (print) | LCC BF448 (ebook) |
    DDC 170—dc23/eng/20220517
LC record available at https://lccn.loc.gov/2021060558
LC ebook record available at https://lccn.loc.gov/2021060559

10 9 8 7 6 5 4 3 2 1

Dedicated to all of my colleagues over the years in so many different fields who have worked with me, shared their knowledge with me, and welcomed me into their communities. I am better for having taken this journey with you.

We all do better when we all do better.
—Paul Wellstone

Won't you be my neighbor?
—Fred Rogers

Contents

INTRODUCTION

1 SEARCHING FOR A SCIENCE OF MORALITY

About ten years ago, I was flying on a plane back to my home in Minneapolis from some scientific conference and found myself sitting next to a pastor returning from a mission trip with his church. We got to talking, and the pastor and I had a fascinating conversation about science and morality.

My own research lies at the boundary between neuroscience and economics and studies how humans and other animals make decisions. My colleagues and I had discovered that people possess multiple decision-making systems, each of which uses information about the world in a different way. In a sense, humans are not unitary beings—we are multiple selves. And that fact explains a lot about how and why we sometimes find ourselves being inconsistent and taking actions that surprise us.

Some of our decisions depend on an explicit consideration of potential options and outcomes (called the *deliberative* decision-making system), while other decisions are made only after a lot of repeated practice (habits, also called the *procedural* decision-making system), and still other decisions are ingrained (species-specific, species-important) behaviors that we know how to do but learn when to release (we call these *Pavlovian* decision-making, but "instinctual" is probably a better word). Running away from a lion is Pavlovian, while hitting a baseball is procedural, and deciding where to go to college is deliberative. But it turns out that laughing with your friends is also Pavlovian, as is responding to unfairness.

My colleagues in this field of neuroeconomics had been studying economic games—"Do you share your resources with another?" "If another player in the game doesn't share with you, what do you do?" "If another player in the game cheats a third player, do you spend your own resources to punish them?" "Should you kill one person to save five?" In a very real sense, these economic games asked moral questions—and my colleagues had made two important discoveries. First, many moral decisions depend on the Pavlovian system, not the deliberative,[1] contrary to almost everyone's expectations and centuries of Western philosophy on morality. And second, one of the few differences between humans and other animals, particularly other mammals, and particularly other primates, was growing evidence that humans might be handling these moral questions differently.[2] All of this meant that science was starting to get a handle on ways to study morality and to ask questions scientifically about it.

And I asked him, this pastor I had met on the plane, what he would do if (I think I said "when") science is able to explain morality. He said he wasn't worried because he had faith that science would find that what it took to be a moral person was what religion had been arguing all along—that we are all part of a community and that there are right and wrong behaviors within that community, based on helping others.

I had no idea how right he was.

WHERE THIS BOOK COMES FROM

My own research lies in the neural mechanisms of decision-making, looking at how we process information about the past (memory), present (perception), and future (potential outcomes and consequences) to take actions. We now know a great deal about how those decision systems make the choices they do, and we have found that how you ask someone to do something changes how each decision system responds and, thus, how that person responds to the question. Moreover, because each decision system processes information differently, how you ask someone to do something can even change which system drives the behavior.

Much of this research lies in the burgeoning field of neuroeconomics. Over the last decades, neuroeconomic experiments have discovered that these different decision processes influence moral decisions in interesting ways and interact with how humans cooperate. In particular, researchers have discovered that certain decision components include intrinsic goals of fairness and cooperation.

In step with this, over the course of the last decade I have been finding my research relating more and more closely to these questions and making connections to the literature on altruism, community, and third-party punishment such as that by Elinor Ostrom, Ernst Fehr, and David Sloan Wilson; to the neuroeconomic work examining the neuroscience underlying behavior in variants of economic games by researchers such as Josh Greene, Molly Crockett, and Read Montague; to the evolutionary and anthropological data from scientists such as Jane Goodall, Frans de Waal, Chris Boehm, and Michael Tomasello; and even to the moral philosophy of John Rawls and Tim Scanlon. I have to admit, I discovered Scanlon's work (e.g., *What We Owe to Each Other*) through the amazingly deep TV comedy *The Good Place*, which is all about how selfish individuals (first Eleanor and then Michael) discover their morality through the importance of community ("Eleanor, find Chidi").

But these literatures still felt disconnected, particularly when the neuroscientific integrations fell back on trivial philosophies of morality in their assumptions about the goals of morality (like *utilitarianism*, the belief that the optimal decision is the one that makes the most people the most happy) that clearly didn't fit the actual behavior of humans—humans simply do not behave as if they are utilitarians.[3]

Many of the sociobiology books I was reading at the time claimed that morality is an epiphenomenon of calculations made by selfish people and that altruism and community are the mistaken consequences of genes trying to propagate their own kin.[4] These ideas, while good descriptions of ants and bees,[5] didn't fit the behaviors we were seeing in the neuroscience and economics literature. Humans were doing something more complicated than these simple selfish-gene models were arguing for. Something in that puzzle was missing.

I'd been feeling for a while that there was an integrated breakthrough that could be found here, bringing together all of these different literatures, but I didn't see it until I started collaborating with my colleague C. Ford Runge, who pointed out that we had been looking at *ultimatum games* and *prisoner's dilemmas* when the real key was a different game, the *assurance game*. I will describe these games in depth in chapters 2 and 3, but quickly stated, in these games one decides whether to cooperate with a partner or to cheat them. In the single-interaction prisoner's dilemma game, it is always better to cheat, although humans often cooperate anyway. In the assurance game, however, it's better to cooperate with your colleagues if they are going to cooperate with you—but not if they aren't. Interested readers who can't wait for the details can also find summaries in the glossary.

This insight answers one of the most difficult questions in moral philosophy. Much of moral philosophy is about doing the right thing *despite* the fact that it's bad for you as an individual. But if we are in an assurance game, not a prisoner's dilemma, then being moral is not only good for the community, it is good for the individual as well. Thinking of humanity as playing an assurance game could explain why a community-oriented species could evolve, but something was still missing.

Moral statements always seemed to be about a lot of other things than just cooperation. Certainly, some moral codes, such as those laid out in many religions and philosophies, do address questions of cooperation, including exhortations to work with the group and punishment for transgressions against communities. They include beliefs about the importance of autonomy and about protecting the innocent. But religions also have beliefs and rules that go beyond cooperation. Similarly, studies of what humans identify as moral and immoral, such as Jonathan Haidt's enumeration in his 2012 book *The Righteous Mind*, include admonishments and exhortations beyond group cohesion. Haidt argues that morality includes a great deal more than simple questions of equality and fairness. How does "purity" or "disgust" relate to questions of cooperation?

And then I found the recent evolutionary psychology book by Michael Tomasello, which talked about evolved mechanisms that encourage cooperation in an assurance game as "Morality." I realized that we could extend these ideas from evolved intrinsic goals to general moral codes. Moral codes were a toolkit to help people be good. Social structures are technologies; just like the wheel or nuclear power, they change how we interact with each other.

I realized I had seen this idea of social structures as technologies before, in the strategy-based *Civilization* games that I have often played with my kids. In these games, discovering technologies provides your civilization with new abilities, whether the discoveries are physical, like the wheel (providing roads) or nuclear fission (revealing uranium, providing nuclear power plants and nuclear bombs), or social, like currency (enabling economic controls), theocracy (changing how religion spreads in your and other civilizations), or democracy (changing corruption levels in your civilization). Of course, the *Civilization* series and its ilk are games, not simulations, so we don't want to take any specific consequences of a social technology as a scientific hypothesis. Nevertheless, the primary insight that social structures are a technology like any other remains true. To be fair, once I realized this, I was able to track down similar insights well established within the philosophical, sociological, and economics communities. Social institutions change the rules of the game.

And it all fell into place. Suddenly, all of those literatures made sense to me. The social structures of community as observed in anthropological and socioeconomic studies and the moral structures enumerated by people like Jonathan Haidt or the extensive discussions of what one should or should not do in moral philosophy are the tools we use to enable us to achieve moral goals; they are not the moral goals themselves. Crucially, this insight avoids the pernicious problem in moral philosophy of *moral relativism* (the idea that all moral structures are equally valid). There really is a moral goal—that of converting the game into an assurance game. (In an assurance game, cooperating is best for *both* the individual and the group, making it more likely that the individual will cooperate.) And most importantly, some tools work better than others.

Toolmaking is a form of engineering, and engineering depends on science. In order to understand these tools, I needed three scientific literatures. First, I needed scientific studies of how small groups of people interact when faced with moral questions (*mesoeconomics*). Second, I needed to know how changing the rules changed the game (*sociology*). And third, because those rules interact with our decision-making systems, I needed to know how those decision-making systems worked and how they interacted with our environments, including our social environments (*neuroscience*). I realized that pulling these three literatures together could give us a new science of morality.

Science is a process. It assumes a reality that can be understood through explanations of phenomena. Sometimes those explanations come from generalizing concepts (such as the way we explain the foaming that happens when you mix baking soda and vinegar as an interaction of bases and acids). Sometimes those explanations come from interactions with other objects (such as the way we explain that leaves are green because they reflect the color of the light that is not absorbed when chlorophyll transforms photons into a plant's chemical energy stores). And sometimes those explanations come from predictions about optimal outcomes and the deviation from those optimal outcomes (such as the way we can say that human eyes are imperfect because the light-sensing cells lie below the information-processing cells, which explains the problem of the blind spot, where the wires from the information-processing cells travel through to the rest of the brain).

Generally, the process of science is that we identify predictions from a theory within a framework and try to explain strangeness in the data from those theories. For example, in the nineteenth century, careful observations of the motions of the planets fit our understanding of physics from Isaac Newton and Johannes Kepler very accurately, except for Uranus and Mercury (the farthest outward and inward planets known at the time). Within that Newtonian/Keplerian framework, one could explain the weirdnesses in the planetary orbits by positing two additional planets, one pulling slightly on Uranus and one on Mercury. In 1846, the prediction of a planet beyond Uranus came through with amazing success, identifying the planet Neptune exactly as predicted. However, attempts to find a planet even closer to the sun than Mercury did not succeed. But when Albert Einstein changed the Newtonian framework, Mercury's orbital deviations fell out beautifully.[6]

What I aim to provide in this book is a new conceptual framework that provides a new definition of morality. This new framework is, in a sense, the language with which we should be discussing questions of morality. Because this moral story is an engineering story, it has a goal, and we can ask normative questions about whether our moral structures get us closer to that goal or not. Once one has defined such a goal, one can create a scientific question of how best to optimize toward that goal. For example, when the goal of

Apollo 13 changed from "landing on the moon" to "bringing the astronauts home safely," the scientific questions being asked also changed.[7] I will show that morality is about achieving a specific goal: that of building, maintaining, and encouraging cooperation within assurance games, increasing the likelihood that our compatriots will work with us rather than against us. To do so, we will have to understand how those moral structures interact with our decision-making systems.

What we do with that scientific understanding can be both good and bad—an understanding of nuclear power has given us both radiation therapy for cancer treatment and nuclear bombs. Similarly, a scientific understanding of morality can provide us with institutions that make our lives better and institutions that make our lives worse. The science is about understanding the process. But there are also design consequences of our scientific understanding, which means that as we gain a scientific understanding of morality (the descriptive, *is* side of morality), we can better design structures that take advantage of that scientific understanding (to better achieve the prescriptive, *ought* side of morality).

The actual process of science is very slow.[8] It can take decades and even centuries to determine the right answer, but the process does work, in a way that other truth-identification processes really don't. And we can use that process to take control of our world. We can fly in airplanes because we understand the physics of wind (and because we understood enough about chemistry to create engines with the required power). This book is not claiming that we've "solved morality," but rather that we have enough of a handle on it to start studying it scientifically. In a sense, I am arguing that scientists have been studying morality scientifically for the last half century, but in separate literatures that need to be brought together to inform each other. This is the beginning of the story, not the end of it.

As Stuart Fierstein points out in his book *Ignorance*, science is about new questions as much as new answers. Einstein's insight didn't just settle the issue of Mercury's orbit; it opened a host of new questions that we have spent the last hundred years examining. These new questions have led to the discovery of gravity waves, to new views on the universe, and even to the computer I am writing this book on, which depended on understanding questions of information and the precision of atomic clocks. I hope that this new framework provides a new set of questions to answer, such as (but definitely not limited to) *How do moral technologies change which decisions get made? How well do moral technologies change the game being played?*

Descriptively, we can study morality by building on the neuroscience and psychology of decision-making and how those decision-making processes interact with moral questions. Importantly, understanding how the world works descriptively has implications for how technologies interact with the world. Knowing about gravity tells us that jumping off a cliff is a bad idea. Knowing how airflow over wings can provide upward lift makes jumping off a cliff with an appropriately built hang glider a viable possibility. Many neuroscientific descriptions of morality have stopped there, which, unfortunately, leaves us in a world of *moral relativism*. However, normative theories can provide a scientific process toward optimization once we have agreed on what is being optimized. Our new descriptive understanding of morality will be critical as we try to develop technologies to improve our lives.

In this book I argue that what we typically call *morality* are the technologies we have invented that make us more likely to find ourselves cooperating on the positive-sum side of the assurance game. This prescriptive science allows us to escape the trap of moral relativism—some moral structures really are better than others because they do a better job of achieving the prescriptive goal of making us more likely to be playing the assurance game and to get to the mutual cooperation corner of that assurance game. Like all prescriptive science, those technologies depend on the descriptive science underlying them—in this case, of how they interact with our neuropsychology. In order to achieve that prescriptive goal, we have to understand the descriptive science and how those technologies interact with our decision-making system.

WHOSE GOOD ARE WE OPTIMIZING?

One of the largest problems in studies of morality is that while it is very easy to explain how communities form—scientific evidence for the idea that humans have evolved to form communities that compete with each other has been eloquently laid out in dozens of books—researchers are unable to say why or even that some communities are better than others. In the end, these books can explain why Nazis exist, but they cannot say why Nazis are bad. (To be fair, Haidt's *The Righteous Mind* and Hare and Wood's *Survival of the Friendliest* both do say that Nazis are bad but can provide no justification beyond the statement itself. The statement is not actually a consequence of their logic.) Nazis are bad because their tools work poorly relative to other options—there are better tools for solving community problems than genocide. In her recent book *Caste*, Isabel Wilkerson provides an eloquent enumeration of how these misstructured moral codes work poorly for both the dominant and subordinate castes. Similarly, while some researchers have identified ways in which one group dominating others can evolve, this new logic can explain why such a group domination process is harmful. Moral codes that encourage dominance may be transiently stable, but in the long run, they prevent us from reaching that cooperating assurance game and reaping the benefits of that cooperation. They make the world worse for everyone (including the dominating subgroup) than those built with more inclusive moral codes.

In this book I am putting forward the hypothesis that morality is about transforming our interactions into an assurance game, in which what is better for the community is also better for the individual. Many books suggest that there is a conflict between the community and the individual; this book argues that well-structured social constructions improve the community for all individuals and that in an assurance game, cooperating helps everyone. To quote the late senator Paul Wellstone, "We all do better when we all do better."[9]

The key insight is that moral codes are actually trying to change the rules of the game to make it more likely that someone will participate in the community and more likely that participating in that community will make that individual's own life better. This means that moral codes are a tool, and there can be good moral codes (that work well) and bad moral codes (that work poorly or have dangerous side effects). Of course, the ability to predict how (and whether) moral codes work will depend on understanding how human decision-making works. Moral codes need to interact with the human decision-making process to help us toe the right path. We will see throughout the book that these codes go way beyond simply rewarding cooperation and punishing selfishness. In fact, sometimes simple monetary punishments can backfire, reducing cooperation.[10]

The answers to these questions are not trivial. They contain complex interactions that require us to work through an understanding of both what the optimal solutions are (the *ought* part of morality) and how humans interact with these social technologies (the *is* part of morality). But we can build from all of the work on the mathematics of economic games, the mathematics of communities and evolutionary biology, and the neuroscience and psychology of decision-making, particularly in the light of moral situations, as well as from works of moral philosophy and legal theory, all of which contain incomplete pieces of the puzzle.

There is an integrated story here. Furthermore, I found that this integrated story fits with the literary and artistic works that have been observing humanity for millennia, which helped to convince me I was on the right track. (Great art is about the human condition. We are very quick to reject art that gets the human condition wrong.)

The pastor was right—we are all part of communities; there is a right and a wrong behavior within communities, based on helping others; and we call those behaviors "morality." Importantly, this also means that not all communities are equal—some are better than others. How we construct those communities can make it easier or harder to stay on the right path, to do the right thing and not the wrong one.

This book is that falling into place—an argument that we now have *a new science of morality* and that, like all sciences, it has consequences for our institutions, our policies, and our understanding of ourselves.

* * *

THE JOURNEY OF THIS BOOK

If you're the kind of reader who skips to the end of a novel to find out what happens, you can jump to the last section of the book (chapters 18–20), where I will lay out that new science of morality, including what it means for building societies. But like reading only the end of a novel, you'll know if the hero survives, but not how or why. To get ourselves to that new science of morality, we're going to need to take a journey to get there. We're going to need three puzzle pieces, which will form the first three sections of the book.

First (part I, chapters 2–6), we need to find a language we can use to talk about our interactions with each other. We will find this in the language of *game theory*, where we create simple games to play. We can ask both *How do humans play those games?* and *How should humans play those games?* (Those are potentially different questions.) We will find, critically, that changing the rules of the game changes the answers to both of those questions.

Second (part II, chapters 7–11), we need to understand how humans make decisions, particularly when faced with moral quandaries. We will find this in the field of neuroscience, and critically, we will discover that the different decision-making systems that make us who we are interact with these morality situations in different ways. We will find, critically, that changing the question changes how we react to it.

And third (part III, chapters 13–17), these two discoveries mean that we can change how we choose by changing the rules of the game, by changing how we ask ourselves the questions of our lives, by changing the society we live in. We will find these answers in the field of sociology, and we will find, critically, that we can make the world a better—or a worse—place for us as individuals by changing these institutions and policies.

These three insights give us the supports on which we can build our new science of morality. If you like, you can skip all of them and jump to the end to find out if our heroes make it home, but the key to understanding the science, like any good story, is in the journey itself. Hopefully, you will find it as interesting a journey as I have. So let's follow our camera as it opens on what appear to be surprisingly simple economic games . . .

I MORAL EXPERIMENTS

2 ECONOMIC GAMES

We explore a way of understanding interactions between individuals through the light of economic games and see that changing the rules can change the choices people make.

I'm going to argue that moral questions are those that depend on our interactions with each other. (I will save questions of our interactions beyond humanity, such as with nonhuman animals, potentially sentient beings such as aliens or robots, and other artificially intelligent agents, for the end of the book, after we've had a chance to derive the logic of morality in the human context. Similarly, I will address questions of intergenerational morality, such as climate change, after we've had a chance to derive the logic of morality within a known interacting group.) To study how humans interact with each other, we can create moral experiments as games to play. These games are simplifications of the complex world, simulations if you will, that allow us to get at the fundamental questions of our interactions.

Readers may be concerned that these kinds of economic games assume that humans (and other agents) always make the choice that is "best for them." This is often referred to disparagingly as *Homo economicus*, the idea that humans maximize the money they can get, but actually, these ideas rest on the assumption that humans maximize an internal subjective value that can include a whole host of factors, such as fairness, reputation, social affiliations, and so on. Furthermore, what gets identified as the best option depends on the specific information processing that goes on in the human brain. As we will see in chapter 7, what one determines to be the "best option" changes based on how a decision process uses its limited resources to calculate its options. All of these issues will play important roles in how people play these games.

Moreover, it is important to point out that we are not going to depend on assumptions of *Homo economicus*. While we will often specify these choices in terms of numbers or money, we do this for convenience of explanation—the fact that "subjective value" is subjective means that, in reality, the reward is not simply "the most money." For example, when we talk about the *tax game* at the end of the next chapter, we will discuss it in terms of "money doubling in the central pot," but that is just a way of simulating things that are cheaper or easier to do as a group, such as building a flood levee to protect all the families in a village.

Let's start with a simple game known as the *matching pennies* game, so called because one version has each person place a penny on the table before them and cover the penny with their thumb. Players then reveal their choices simultaneously. We have two players A and B, each of whom gets to choose an option (we'll call the two options *heads* and *tails*). One player wins if both players choose the same option (both heads or both tails), while the other wins if the choices are different (one chooses heads, the other tails).

Real-life examples of the matching pennies game occur when one player tries to deceive the other, such as the D-Day invasion of Europe in World War II. In 1944, the Germans could prevent an invasion of Europe by the Allies if they concentrated their defenses in the right place. If the Germans guessed correctly about where the Allies would invade, the Germans would prevent the Allies from creating a beachhead, but if they guessed wrong, the Allies would establish a beachhead and retake Europe. There were two locations that the Allies were likely to try: Calais and Normandy. Both were in France just across the English Channel, allowing the Allies to stage a massive army in England in preparation for the invasion. We can write this game as a table. The Germans make a decision about where to defend by selecting a column, and the Allies make a decision by selecting a row. The outcome can be read as the intersection of those two decisions.

Table 2.1		
The D-Day invasion as a matching pennies game		
	Germany	
	Defend Normandy	**Defend Calais**
Allied forces — **Invade Normandy**	**Germans** hold **Allied** invasion fails	**Germans** fail **Allied** invasion succeeds
Allied forces — **Invade Calais**	**Germans** fail **Allied** invasion succeeds	**Germans** hold **Allied** invasion fails

We can see that this is the matching pennies game. If the Germans guess correctly (match), then they hold Europe, but if they guess incorrectly (don't match), then the Allies gain a beachhead in France. As we all know, the invasion of Normandy succeeded, in part because the Germans concentrated their forces at Calais, putting the world into the upper-right corner of the table. This was in large part because the Allies successfully deceived the Germans through Operation Fortitude, which included rubber blow-up tanks and fake troop bases and, most importantly, General George S. Patton, whom the Germans did not believe would be sidelined from the real invasion.[1] Of course, it would be nice if we could study moral questions in a space where we don't need armies of soldiers invading territories, destroying lands, and killing each other. War is a form of a team game, and like war, most team games—football, baseball, soccer, e-sports like League of Legends—have winners and losers. One team wins; the other loses. Maybe they come to a tie, but there is no process whereby both teams win. (In real wars, often, both sides lose, which makes not fighting wars a form of the assurance game, as described in chapter 3.)

When we think of games, we often think of competitive games like chess or tic-tac-toe, where one player wins and the other loses. These games are called *zero-sum games*. If someone wins, someone else has to lose. But not all games are zero-sum. Sometimes it is possible for both players to win or both players to lose. Both real wars and sports games contain a subtlety that goes beyond the trivial statements of zero-sum economics.

First, team games contain two levels of interaction: between-team interactions, which are often zero-sum with winners and losers, and within-team interactions, which are the opposite of zero-sum. All the players within the team win or lose together. At the end of the Super Bowl (the championship game for the American National Football League), every member of the winning team gets a ring, whether they were on the field or not. Certainly, consequences outside the game mean there is a big difference between being the backup who sat on the bench waiting in case something went wrong and being the wide receiver who caught the game-winning touchdown. But within the game, the team wins or loses as a team. Within the game, the backup player ready to step in and save the day when the first-string quarterback goes down is also playing an important role.

Second, a lot of these games have a meaning beyond the within-game events. In sports, we want a "good game," with exciting plays. Sports are often almost a form of dance in which the two sides are creating art together. The best games are those in which the two sides are more evenly matched than expected, and the underdog plays the superstar to the very edge and maybe even wins. This makes for a "good story" and "great drama" and is what people tune in for. Think of the wonderful interaction between Bianca Andreescu and her idol Serena Williams when Andreescu beat Williams at the 2019 US Open, or the famous scene at the end of the original *Rocky* movie where Rocky loses but remains standing and gains the respect of both the crowd and his opponent, Apollo Creed.

And, of course, the key to sports is that there are rules that we follow. Both sides agree to a set of rules that will make the game fair. It is important to recognize that this does not have to happen and that playing by the rules is a metagame in itself. For example, the famous iocaine powder scene in the movie *The Princess Bride* is supposed to be a classic zero-sum game. Westley (a farm boy who has returned secretly as the Dread Pirate Roberts) is supposed to put iocaine powder (in the movie, an odorless, tasteless, fatal poison) into one of two cups, and then Vizzini is supposed to pick which cup each player drinks from. We can write this as a classic zero-sum table:

Table 2.2
The iocaine powder game as Vizzini thinks it is being played

		Westley	
		Poison in cup 1	Poison in cup 2
Vizzini	Drink from cup 1	**Vizzini** dies **Westley** lives	**Vizzini** lives **Westley** dies
	Drink from cup 2	**Vizzini** lives **Westley** dies	**Vizzini** dies **Westley** lives

Notice that this is a matching pennies game, just like our examination of D-Day, above. However, unbeknownst to Vizzini, Westley has built up a tolerance to iocaine powder, and it will no longer kill him. Therefore, Westley has put poison in both cups. This is now a new game:

Table 2.3
The real iocaine powder game that Westley is playing

		Westley
		Poison in both cups
Vizzini	Drink from cup 1	**Vizzini** dies **Westley** lives
	Drink from cup 2	**Vizzini** dies **Westley** lives

Westley has changed the game. This is important to defining his character—the quiet farm boy has become hard and capable of killing, without compunction or concern. We also see Westley change the zero-sum game in a positive way when he knocks out Inigo Montoya after the masterful sword fight in which the two of them realize that they are not competing in a zero-sum game but have instead been dancing together (notice the point made about sports earlier). Their recognition of each other's mastery of sword fighting and their respect for the rules of the game they are playing leaves open the opportunity for them to become friends. As Westley says upon defeating Montoya: "I would sooner destroy a stained glass window than an artist like yourself."

The key to our story is that most of our interactions with each other are not zero-sum games. More often, we are like Westley and Inigo Montoya—trying to learn how to become friends. In a non-zero-sum game, it is possible for both players to win or for both players to lose.

THE PRISONER'S DILEMMA

The first example usually laid out in these stories of economic games is the prisoner's dilemma, in large part for historical reasons because it was the first to be mathematically examined. We will start with it because understanding the prisoner's dilemma will set up a good contrast for why the assurance game is so different.

In the classic (negative) version of the prisoner's dilemma, we tell a story about two criminals caught stealing state secrets by the police. The police have enough evidence to convict them on a minor charge (breaking and entering) but not enough for the major charge (treason). If the police can get one criminal to confess, then they'll make a deal with that criminal, reducing the prison sentence even further while throwing the book at the other criminal. If they both confess, then the police can use the confessions against them but will reduce the sentence as a reward for not having to go to trial. This leads to the following game table. As before, Prisoner A chooses a column, and Prisoner B chooses a row.

		Criminal A	
		Stay silent	**Confess**
Criminal B	**Stay silent**	A gets two years in jail B gets two years in jail	A gets one year in jail B gets life in jail
	Confess	A gets life in jail B gets one year in jail	A gets ten years in jail B gets ten years in jail

Table 2.4
The original (negative version) of the prisoner's dilemma

Importantly, in this game the two prisoners are not able to communicate with each other. Thus, they have to make their choices before they know what the other is going to do. However, working this game out logically, it doesn't matter what the other person chooses. It's always better to confess.

Let's look at this from Prisoner B's point of view. If A stays silent, B can cut a year off their time in jail by confessing, while if A confesses, then B should confess as well to prevent ending up serving a life sentence in prison. No matter what A does, it's better for B to confess. The problem is that an equivalent logic works for A as well—no matter what B does, it's better for A to confess. So both players, playing logically, should confess no matter what, and as a result they find themselves in the confess-confess lower-right corner of the table, serving ten years in jail each, which is worse than if they had both stayed silent. If they had both stayed silent, they would each have served eight fewer years in jail.

I have always found the negative version of this prisoner's dilemma confusing because the term for staying silent is *cooperating*, while the term for confessing is *defecting*. (These terms arise because the criminals are cooperating with each other and defecting against each other. But the criminals defect against each other by cooperating with the police and cooperate with each other by not cooperating with the police, which . . . oy!) So I like to rethink the prisoner's dilemma in a positive light. Let's imagine the following contrived game:

Two players, A and B, get two choices C or D (cooperate or defect). If they both cooperate, they get a small win. If one defects while the other cooperates, the defector gets a big win. If they both defect, they get very little. (The specifics of how much money they are playing for is arbitrary as long as the game has certain properties—that defecting is worth more than cooperating for each individual and that both cooperating is worth more than both defecting, which is what makes this a prisoner's dilemma game.)

Table 2.5		
The monetary (positive) version of the prisoner's dilemma		
	Player A	
	Cooperate (C)	**Defect (D)**
Cooperate (C)	**A** and **B** each get $6	**A** gets $10 **B** gets nothing
Defect (D)	**A** gets nothing **B** gets $10	**A** and **B** each get $1

(Row labels under "Player B": Cooperate (C), Defect (D).)

Note that player A is again better off defecting no matter what player B does. If B cooperates, A can get $4 more by defecting, and if B defects, then A must defect to get just that $1; otherwise, A gets nothing. B's decisions follow a similar logic based on what A chooses. So, again, no matter what, it's better to defect, and if both players play "logically," then they will always defect and end up in the lower-right corner where each person gets only $1.

SPLIT OR STEAL: REAL-LIFE EXAMPLES

One of my favorite experiments looking at human interaction in a prisoner's dilemma game is the British TV game show *Golden Balls*, in which a pair of contestants play a game called *split or steal*. They first play a set of rounds to build up a pot of money and then play one round of the prisoner's dilemma. They each get two golden balls (catch where the name of the show comes from?), one of which when opened reveals "split" and the other "steal." If they both pick split, then they split the pot fifty-fifty. If one picks steal and the other split, the one who picked steal gets all the money. And if they both pick steal, they both get nothing. This creates the following game:

Table 2.6

The *Golden Balls* game as a prisoner's dilemma

		Player A	
		Split	Steal
Player B	**Split**	**A** and **B** each get half the money	**A** gets all the money **B** gets nothing
	Steal	**A** gets nothing **B** gets all the money	No one gets anything

This payout matrix is a form of the prisoner's dilemma. If your opponent splits, it's better economically to steal. If your opponent steals, then, economically, it doesn't matter to your income whether you split or steal, but choosing split marks you as a sucker on national television (international television). Interestingly, people defect more on the show than they do in laboratory experiments.[2] While no one actually knows why, one possibility is that they are on television, and no one wants to appear to be a sucker on international television, which suggests that there is a meta-game going on here. We'll explore that meta-game when we come to questions of reputation and social interaction in a few chapters. For now, I want to use this split-or-steal game to help us understand how humans play non-zero-sum games.

Importantly, *Golden Balls* includes a chance for the two players to talk to each other and to convince each other to choose split or steal, usually by claiming that they themselves are going to split. (Sometimes they are lying.) However, the choices are revealed simultaneously, so players must still make their decision without knowing what the other player has actually decided to do.

There are lots of examples of the split-or-steal scenes from *Golden Balls* on YouTube and other media (it makes great television). In one video, Sarah and Steven play for £100,000. (It's a British game show, so the money is in pounds.) During the conversation, Sarah says her friends would be disgusted if she chose steal; Steven says he'd be happy to go home with £50,000. When he says that, she closes her eyes and makes a sad face, and everyone not in the moment knows exactly what's going to happen. She is going to take all of the money. And she does. In the moment when they discover what has happened, he hangs his head with his hands over his head, and she looks to the audience with a satisfied smirk, knowing exactly how she played him. She does not apologize.

It is clear that it cost Steven something to lose all that money. This suggests that fairness is an important part of human behavior.

FINDING A WAY TO FAIRNESS

It is clear from the split-or-steal example scenes that fairness is a key factor in human decisions. Both the audience and many of the players seem to have the goal of finding a way to split the money. I think this goes beyond not wanting to look like Steven (i.e., to be a loser, to look like a chump).

In a second example, two women, Michelle and Ella, play for £30,000. In their discussion, Michelle says that because of her religion she would never steal. She says that she's going to give her half of the money to her church. Ella acknowledges Michelle's goals, smiles, and says, "Let's go home happy people." In the end, Michelle chooses to split, and Ella chooses to steal. Afterward, Michelle declares that she would have split no matter what because that's what she felt was the right thing to do, while Ella says, "It's just a game, and I won." I find this example interesting for a number of reasons.

First, Michelle and Ella seem to be playing different games. Michelle is disappointed in Ella and clearly thinks Ella failed to do the right thing. Ella is triumphant. Second, Michelle justifies her decision through religion, which, as we will see in chapter 15, plays a very interesting role in understanding morality. But, third, a careful viewing of the scene finds that Ella never lies. It's as if she has her own moral line she is unwilling to cross—she will pick steal but won't lie about it. She never says she's going to split. She says she's happy with her choice. She acknowledges Michelle's game play. To Ella, the game is about winning through fair play and she won. To Michelle, the game is about finding a way to work together, and they both lost. Ella and Michelle are playing different games.

In a third example, two men, Ashleigh and Ben, play for £35,000. They have both lost in previous rounds and are unhappy about it. In their discussion, they acknowledge how it felt to lose the game, and they talk about how they would like to redeem each other. In the end, they both pick split, and the entire audience celebrates. The two men hug each other in relief. They escaped the lower-right £0/£0 corner, and both men are happy to have found a way, together, to get to the upper-left split £17,500/£17,500 corner.

If you look for split-or-steal scenes from *Golden Balls* on YouTube, the most-viewed scene is a remarkable case in which Ibrahim and Nick play for £13,600. When they start their discussion, Nick says, "You have to trust me 100 percent. I'm going to steal. If you split, then I will give you half of the money after the game." Everyone is shocked. The moderator points out that Nick has no obligation to share the money after the game. Ibrahim says, "That's not how the game is played!" They argue and Nick holds his ground. By the end of the scene (which was apparently cut down from over forty-five minutes of argument[3]), Ibrahim has clearly given in and really believes Nick is going to steal. So his only option is to play split and hope that Nick will hold up his end of the bargain (which Nick has announced publicly in a television broadcast with a permanent record). In the end Nick plays split, and the two of them go home sharing half the pot. What Nick has done is changed the game (like Westley in the iocaine powder scene). He has ensured that he knows what Ibrahim is going to choose, allowing him to decide to play split safely (so they can end up like Ashleigh and Ben and not like Steven or Michelle).

First, let's look at what Nick said (from Ibrahim's point of view):

Table 2.7

Ibrahim's view of the split-or-steal game after Nick's declaration that he is going to steal

		Nick	
		~~Split~~	Steal
Ibrahim	Split	~~Nick and Ibrahim each get half the money~~	**Nick** gets all the money but will give half to **Ibrahim** after the game
	Steal	~~Nick gets nothing~~ ~~Ibrahim gets all the money~~	No one gets anything

Ibrahim has no real choice here. If he steals, then he knows that neither of them gets anything. If he splits, then he has a chance of getting half the money. Importantly, Nick has said publicly that he will share the money, and presumably, Ibrahim could shame Nick into giving him the money after the game, even though Nick has no legal obligation to do so. (We will talk about shame and reputation later, but notice the moral words being used here.) So Ibrahim is going to split. Now that we know that Ibrahim is going to split, Nick's table looks different:

Table 2.8

Nick's view once he knows that Ibrahim is going to split

		Nick	
		Split	Steal
Ibrahim	Split	**Nick** and **Ibrahim** each get half the money	**Nick** gets all the money but will give half to **Ibrahim** after the game
	Steal	~~Nick gets nothing~~ ~~Ibrahim gets all the money~~	~~No one gets anything~~

Nick's choice is obvious: choose split and share the money now—which is what he does. (Because Nick would have to pay taxes on his winnings before sharing with Ibrahim, it is in fact economically advantageous to simply split first.) What Nick realized was that the real key to the split-or-steal portion of *Golden Balls* was to be sure that Ibrahim was going to choose split.

In a later interview on the show *RadioLab*,[4] Ibrahim says that he was originally going to steal and that all of the things he said to justify his moral desire to split were lies. In the interview, Ibrahim laughs and says that Nick successfully conned him into winning almost £7000, which was better than the £0 he expected to get by being sure that he was not going to be played for a sucker.

What Nick did was change the game so that it was no longer a prisoner's dilemma. He found a way to ensure that he could choose split safely because he knew it would put them into the split-split upper-left corner of the table. We call this game an assurance game.

3 THE ASSURANCE GAME

In an assurance game, working together is actually better than working apart, if only we can be sure that our compatriots are going to work together with us.

In an assurance game, both players cooperating leads to a big advantage—but only if both players cooperate. The classic assurance game was introduced by Jean-Jacques Rousseau and is known as the stag hunt. Imagine that we have two hunters in a tribe trying to find food. Individually, they are both good enough at hunting to catch a rabbit, but if they work together, they can catch a deer (a stag). The problem is that individually, hunting the deer will fail. So if they cooperate, they each get half a deer (a lot of valuable meat), but a hunter who cooperates while the other defects gets nothing.

| Table 3.1 | | | |
The stag hunt game			
		Hunter A	
		Hunt deer (cooperate)	Hunt rabbits (defect)
Hunter B	Hunt deer (cooperate)	Hunters **A** and **B** split the deer (much food for each)	**A** gets a rabbit (some food) **B** gets nothing (starves)
	Hunt rabbits (defect)	**A** gets nothing (starves) **B** gets a rabbit (some food)	Both hunters get rabbits (some food for each)

This description was derived from the anthropological literature and the study of hunter-gatherer groups using Stone Age technologies, where it really does take several individuals to bring down a large prey, and because there is no freezer technology to store that prey, it must be eaten immediately—enforcing sharing within the group. (If we all share the deer I get today, then we can also all share the deer you bring in tomorrow.) Since the stag hunt doesn't really make sense given modern technology (individuals can hunt deer successfully and can freeze and store meat for the future, and moreover, most people are not hunting deer for their basic sustenance), let's see if we can find another example.

A great example is infrastructure. Let's imagine that we are the mayors of two towns separated by a river and want to build a bridge so that we can trade. We know that if we build the bridge, it will benefit both our towns. But if Town A spends their money on the bridge and Town B does not, then Town A is left with a useless half a bridge leading out into the middle of the river. A similar logic applies to Town B. I only want to spend my money on my half of the bridge if you are also going to build your half.

Table 3.2

The infrastructure frame of the assurance game

		Town A	
		Build half a bridge (cooperate)	**Save money (defect)**
Town B	**Build half a bridge (cooperate)**	There is a working bridge between the two towns	More money for Town **A** **B** built a bridge to nowhere
	Save money (defect)	**A** built a bridge to nowhere More money for Town **B**	Both towns saved some money

Let's break this logic apart carefully. In an assurance game, if the other player cooperates, then it is better for you to cooperate than to defect. If the other player defects, then it is better for you to defect than to cooperate. In an assurance game, the best option is to do what the other player does—cooperate if they cooperate, defect if they defect. Making the right decision in an assurance game requires predicting what the other player is going to do.

This makes the assurance game fundamentally different from both the zero-sum and prisoner's dilemma games. The game is not zero-sum because it is possible for both players to win big (if they both cooperate). In the prisoner's dilemma, when the other player cooperates, it is mathematically better to cheat (defect). This is what makes the prisoner's dilemma a "difficult choice" (a dilemma)—individually, it is better to defect, but both players defecting is worse than both players cooperating, creating a conflict between the individual and group optimal outcomes. But in the assurance game, if you are absolutely certain that the other player is going to cooperate, then it's actually better to cooperate too. Notice, also, that both players cooperating is better than both players defecting. So, while it is better in an assurance game to defect than to cooperate if the other player defects, if you can get the other player to cooperate, then you can both win big.

This changes things completely. Now your goal is not to determine how to cheat the other player without letting yourself get cheated (as we saw with Sarah and Steven or Ella and Michelle); instead you need to figure out how to get your partner to cooperate (as Ashleigh and Ben did). What you really want to do is force the other player to cooperate (as Nick did). We will see in the rest of this book that humans have developed social institutions and social norms that interact with our evolved decision-making processes and learned skill sets to do just that—to increase (and enforce) cooperation in assurance games.

THE PROBLEM OF THE COMMONS IS AN ASSURANCE GAME

One of the classic stories in morality economics is the *problem of the commons*, often described as the *tragedy of the commons*.[1] The problem of the commons is often discussed as a prisoner's dilemma game, but, particularly for renewable resources, the problem of the commons is really an assurance game, especially when played with the same set of compatriots over time.

The original story of the commons is about the difference between what's best for an individual and what's best for the community: Imagine a small town of farmers. Each has their own grasslands to graze their cattle on, but there is also a shared town center, a *commons*, available to all the farmers. As an individual, the first person to use the commons increases their resources relative to everyone else because they get to use up the common resources while keeping their own land as backup. As an individual, the table looks like this:

Table 3.3		
The problem of the commons		
	Farmer A	
	Graze on private land (cooperate)	**Graze on the commons (defect)**
Everyone else — **Graze on private land (cooperate)**	No one is using the commons	Farmer **A** gets the common resource
Everyone else — **Graze on the commons (defect)**	Farmer **A** loses the common resource	Farmer **A** gets some of the common resource

Table 3.3 shows that it is always better for Farmer A to graze cattle on the commons before grazing on their own land. Of course, this is true for all the farmers, so everyone is going to rush to use the common resource, which depletes it. This has led to the term the *tragedy of the commons* because the assumption is that it is a prisoner's dilemma game—each individual is better off using the commons first, so everyone rushes in to use it, destroying it.

The problem of the commons is a general problem of shared resources. Economic analyses of common resources have found that it is important to determine whether a resource is *excludable*, meaning it is possible to prevent people from using that resource (or not), and whether a resource is *rivalrous*, meaning only one person can use the resource at a time (or not). For example, a piece of food is both excludable and rivalrous, while a flood levee barrier is neither. Things that can be bounded but shared (like parks) are excludable but not rivalrous, while renewable resources are generally hard to exclude but rivalrous. These details are beyond the level of this book, but I refer readers to the classic works by Elinor Ostrom, C. Ford Runge, and their colleagues for in-depth descriptions of these different aspects of resource allocation. Here, the key is going to be things that are not excludable whether they are rivalrous or not—if I build that flood levee around our village, you will also benefit from it, even if you didn't spend any of your own money to build it.

So let's come back to the problem of the commons. In the 1990s, the cod fisheries in the North Atlantic collapsed because new technologies made it possible for individual trawlers to obtain huge catches, which led to depleted cod stocks. Again, this is usually described as a form of the prisoner's dilemma—if the others limit their catches, then one trawler could do better than anyone else by not limiting their catch; similarly, if no one else limits their catch, then the one trawler limiting their own catch was only destroying their own livelihood and still not protecting the fishery. The problem is that no one limited the catch, the fishery collapsed, and the Canadian government had to step in with a moratorium on cod fishing. To this day, Canada, Iceland, and other national and international governments impose severe limits on the size of the catch.[2]

But the cod trawlers were not playing this game once. They were playing it for their entire lives, catching cod every year. Moreover, most of them wanted to hand that livelihood down to their children, just as it was handed down to them by their parents. This certainly wasn't a one-time economic game. It was a repeated game, and a repeated prisoner's dilemma is really an assurance game. To see this, let's examine the prisoner's dilemma from chapter 2 (table 2.5), reprinted here:

Table 3.4			
The monetary (positive) version of the prisoner's dilemma			
		Player A	
		Cooperate (C)	Defect (D)
Player B	Cooperate (C)	**A** and **B** each get $6	**A** gets $10 **B** gets nothing
	Defect (D)	**A** gets nothing **B** gets $10	**A** and **B** each get $1

In this game, if we play a bunch of rounds with no expected end horizon and we cooperate every time, we average $6 per round each. If we defect every time, we average $1 per round, which is obviously worse. The important point about the payout matrix in table 3.4 is that if we alternate trials of *I cooperate and you defect* with *I defect and you cooperate*, then we average $5 per round each, which is less than the $6 per round we make by both cooperating every time. If we are playing with the same people over and over (that is, if we are part of a community), then even a game that looks like a prisoner's dilemma on the surface (both cooperating is better than both defecting, but individually defecting is better than cooperating) can actually be an assurance game if we play it repeatedly.

Managing a shared resource is really an assurance game. If everyone else willingly limits their catch, then it is fine for an individual to limit their catch too. Then the supply can be sustained, prices will stay reasonable, and, presumably, the industry will continue for a long time. It's better to be in the cooperate-cooperate upper-left corner of the matrix.

Table 3.5
The problem of the commons

		Trawler A	
		Limit catch (cooperate)	**Overfish (defect)**
Everyone else	**Limit catch (cooperate)**	Stocks remain intact Everyone has a continued livelihood	Trawler **A** gets more fish than other trawlers . . . for a time
	Overfish (defect)	Trawler **A** loses the common resource, and it's worse when the stocks collapse	Trawler **A** gets some of the common resource, but the stocks collapse

So how do we ensure that everyone is limiting their catch?

The most common assumption is that control has to be imposed from the outside, such as by a centralized government agency. This is in fact what ended up happening with the cod fishery. National and international governments imposed severe limits on the catch.

The other commonly described solution is to privatize the resource. If Farmer A owns both the commons and private land, then Farmer A knows they have the commons in reserve, and there is no need to overgraze their land. A well-studied example of this is the *enclosure* system in Britain in which the English nobility privatized common lands under a legal statement concerned about the danger of overuse. Modern studies have found, however, that the enclosure process tended to produce more overuse, not less. Early towns had an excludable commons ensuring that members of the town could share the use of the commons, while other towns could not. Over the course of the sixteenth to the nineteenth century, these commons became more and more privatized, excluding more and more individuals from these resources. Studies of the effect of these enclosure processes have found that they actually increased overproduction and mostly redistributed wealth from the common population to the rich landowners.[3]

Consistent with this idea that restricted but shared resources within a community are generally protected, in a study of lobstermen in Maine, James Acheson found that lobstermen defended lobstering territories and that territories with perimeters were better preserved than territories with porous borders. But he also found that these groups were invested in protecting the long-term viability of the lobster catch. Note that these were groups of lobstermen, living in a hamlet with shared social resources and a shared social structure, not a single avaricious individual. There is something about the resource being owned by a community rather than an individual that we will need to understand as we try to understand the question of morality. As with the original commons shared by the people of those English towns, there are deep differences between groups maintaining a resource and individuals.

Many examples exist where common resources have been successfully preserved, ranging from water resources in Bali to the lobster fisheries in Maine, the sea bass fisheries in Chile, and the halibut fisheries in Alaska. We will see that a particular combination of processes allowed these fisheries to succeed, including a definition of a group with shared ownership, a belief that the individuals were all invested in that shared ownership, and escalating punishments for breaking the rules. These punishments were imposed by the group as a whole and allowed for forgiveness and redemption given sufficient expressions of contrition by rule breakers. Extreme rule breakers were ostracized. The next chapters will show that these processes are exactly the keys needed to solve this generalization of the assurance game.

The common resources problem is a group-level example of an assurance game. Generalizing from our assurance game example of the two towns and the bridge, we can use the question of infrastructure as a good story to look at when trying to understand these problems. I like to call this the *tax game*.

THE TAX GAME

Let's imagine we live in a community. People in this community pay taxes to the government, which then uses those taxes to pay for infrastructure that is useful to the entire community. For example, the government might build a flood levee that protects the entire city. We could ask each individual to put sandbags around their own house, but that would be much more expensive and much less efficient. Importantly, if only some people pay their taxes, while others do not, and the levee gets built, even the houses of people who did not pay their taxes remain safe from the flood. That is, in this game we cannot exclude the people who did not pay their taxes from reaping the benefits of the infrastructure.

As we did when we tried to understand pairs of individuals, let's create an economic game that addresses the problem of these shared resources so that we can study it. These economic games are morality experiments that allow us to both explore the consequences of given choices (*What's the best thing to do given the rules of the game?*) and to measure what actual humans do when faced with different game variations (changing the rules).

I like to call this the tax game, but in the literature it is often called the *public goods* game. I prefer the term *tax game* because I think the language is more evocative. (I recognize that taxes work differently for governments that print their own currency, but such discussions are beyond the scope of this book.[4]) This game is an extension of the two-player prisoner's dilemma and the assurance games we've been investigating but is played by a group of people rather than a pair of individuals.

In the basic version of the game, there is a tension such that what is best for the individual is bad for the group, much like the tension between individual choices and interaction choices in the prisoner's dilemma. (In the prisoner's dilemma, it is better for the individual to defect, but both defecting is bad.) However, like a repeated prisoner's dilemma, a repeated tax game is actually an assurance game. Furthermore, we will find that there are ways to manipulate the game (changing the rules) that make it more likely that individuals will pay their taxes and cooperate with the group.

In the tax game, we have a set of players. Each player has some money, which we call their *endowment*, of, say, $100 each. Each player can put some portion of their money into a central pot. The money in the central pot then doubles and is paid out to everyone equally. The doubling of the central pot is a way of modeling the fact that building one levee is cheaper than everyone individually placing sandbags around their house. We are trying to model situations in which the central infrastructure is more efficient than individual action.

Politically, particularly in the US, there is an ongoing argument about whether private or public institutions are more efficient. Some things are more efficient when done privately, while other things are more efficient when done publicly. At this point in the book, I do not want to get into the question of which is which (we will discuss some examples of each in later sections). I will merely point out that one can do studies of such efficiencies and that the story is not simple on either side. The tax game models situations in which centralized action is more efficient than privatized action. We will talk about this tax game as if it is a simple story of money put into the central pot and divided out. Of course, in reality individuals need to retain some money for their own private purchases. There are big differences between a 90 percent tax rate on a billionaire and a 90 percent tax rate on someone barely scraping by. (The top marginal tax rate in the US in the 1950s was 90 percent.[5]) The simple game we are describing glosses over this issue.

The key to our tax game is that because we are modeling situations in which we cannot exclude use of the resource (such as being protected by a flood levee), the money from the central pot pays out to everyone, no matter how much they put in. So if we have ten players and each person puts in $50, the pot contains $500, which doubles to $1000 and pays everyone $100. Every player goes home with $150 (the half of their endowment they didn't put in plus the $100 they got back from the pot).

Now imagine that one person doesn't put any money in but keeps their $100 to themself. In the economics literature this person is called a *free rider*. So only nine players put $50 in. The pot now contains $450, which doubles to $900 and pays everyone $90. So the nine players who put $50 in have $140 (the half of their endowment they didn't put in plus the $90 they got from the central pot), and the one person who didn't pay their taxes has $190 (the whole endowment they kept for themselves plus the $90 they got from the central pot).

Finally, let's imagine that no one puts any money in. After all, no one wants to be the sucker who is allowing the selfish player to free ride. In this third case, there is no money in the central pot, so there is nothing to double, and everyone has only what they started with ($100). Notice that this third group fails to increase their wealth. The first group of cooperators made a lot more money playing this game. The tax game as laid out here is a generalization of the prisoner's dilemma; there is a conflict between the success of the group as a whole and the individual.

We will call the people who pay into the pot *cooperators* and the people who don't *selfish*. The literature (particularly the exceptional work by David Sloan Wilson) calls team players *altruists*. I'm not fond of that term because it leads to philosophical arguments about whether this is really altruism. If someone takes an action that benefits themself because they are in the cooperate-cooperate corner of the assurance game or because they are playing with a group of cooperators, are they really behaving altruistically? (The famous quote is "Scratch an altruist and watch a hypocrite bleed.") This sentiment, that all altruism is actually self-interest, underlies a host of biological arguments popular through the 1960s to the 1990s.[6] Of course, the mathematics of these two types of players is well defined, so the philosophy is irrelevant. In order to avoid the philosophical baggage that "altruism" brings, I will call them *cooperators*. But the tenet of this book is, in some sense, that the science of morality is about this question of cooperation and selfishism, that humans have developed structures that encourage cooperation, and that these structures are what we call "morality." In my view, whether one wants to call these people altruists or not is irrelevant. What matters is how we can get people to contribute to the central pot in the tax game.

The tax game is a multiplayer prisoner's dilemma. No matter what your compatriots do, it is better to defect (be selfish), but if everyone is selfish, the group ends up with no money in the central pot and no increase in wealth. Thus, just as in the prisoner's dilemma, there is a conflict between the individual-optimal and group-optimal choices. We saw earlier in this chapter that a repeated prisoner's dilemma is actually an assurance game. A similar analysis finds that a repeated tax game has a fundamentally different structure that is also more akin to the assurance game than to the prisoner's dilemma.

Let's imagine that we are playing the tax game over and over. Moreover, let us assume that the resources we have change how quickly we expand. (If these are biological resources, they increase the ability to reproduce. If these are economic resources, they increase the ability of our community to grow.) In our contrived world, we can create new agents to join our game based on the amount of resources each simulated player has; we can clone agents proportionally to the resources of the agent. For simplicity, we will assume that the new agents are members of their parent's group and that they are cooperators if their parent is a cooperator and selfish if their parent is selfish. Notice that within a group, selfish players have more resources, so within a group, selfish agents grow faster than cooperators, but groups of cooperators have more resources and so grow faster than groups of selfish players.

This creates a form of evolutionary *group selection* through individual Darwinian processes. As with the discussion of altruism, there is vigorous debate about whether this multilevel selection theory should really be called group selection or not. Again, in my view the terminology is irrelevant because the mathematics are well defined. What we care about in this book are the consequences of the mathematics.

Analyses of these games find that there are specific factors that encourage group-optimal choices over individual-optimal choices, such as separation and identification of the group and how useful cooperation is over individual action, particularly in contrast to an individual's ability to gain resources. These predictions are able to identify which animals tend to cooperate as groups in biology, even for seasonal cooperators that only cooperate in certain seasons when these properties occur. David Sloan Wilson has argued that this dynamic balance is a general issue when dealing between biological levels in which individuals separate into groups, including genes within an organism, individuals within a society, and societies in the world.[7]

But let's take a deeper look at this dilemma in the tax game. One of the key points of the two-player assurance game is that it is best to do what your opponent is going to do (cooperate if they will but don't cooperate if they won't). Not surprisingly, people's willingness to contribute in the tax game depends on the amount that other people put into the game. In fact, over repeated plays of the tax game, groups tend to dichotomize. Individuals who find themselves in a group of other cooperators tend to increase their contributions (up to a point, which tends to approach about half of their endowment), while individuals who find themselves in a group of selfish players tend to decrease their contributions.[8]

What this means is that playing selfishly (defecting) in a repeated assurance game has consequences. If you play selfishly in a group of cooperators, then you will help drive your community into the no-one-contributes situation, which is bad for you as an individual. Assuming you are playing a continuous game with no expected end horizon, you want to be living in a group of cooperators because you will do better there (as an individual) than in a group of selfish players. The repeated tax game is a form of an assurance game.

So what are the keys to this repeated tax game that make it an assurance game? First, we need to be playing the game over and over again with the same players (we need to be part of a defined community). Second, individual players need to react to the choices made by their compatriots (cooperating in cooperative communities but not in selfish communities). And third, there needs to be some sort of competition between communities (importantly, this competition need not be violent conflict—we only need cooperative communities to do better than selfish ones). Humans have lived in just these sorts of communities for a very long time.

However, I want to argue that this tax game insight doesn't actually get us to the question of morality. It is not enough to say that humans are members of these kinds of communities or that humans have evolved to solve the assurance game. As with the two-player games we looked at earlier, what the tax game gives us is an experimental paradigm we can use to probe questions of morality. We need an additional step to help us actually define morality.

THREE PROBLEMS OF A GROUP ASSURANCE GAME

Extending the assurance game to a community identifies three dangers: the *free rider*, the *enemy within*, and the *enemy at the gates*.

The free-rider problem occurs when an individual takes from the group without giving their fair share (when they cheat the group). To solve the free-rider problem, we will need social constructions to ensure that we are only cooperating with others willing to cooperate with us and that free riders are not able to accumulate individual gains at the expense of the group. Importantly, these social constructions need to prevent two kinds of free riders: crooks and cheaters who lie or deceive and bullies who use threats, force, or other forms of persuasion to disrupt the group. We will see that these two types of free riders require different social norms. Similarly, we will see that we will need *metanorms*, social norms that say, "It is not acceptable to let someone get away with breaking the social norms." This observation will lead to the importance of using coordinated action to control free riders.

The free rider is a loner, and we can create communities to deal with them. But what if the free rider has friends of their own? What if there is a subgroup that cooperates among themselves but decides they no longer want to cooperate with the rest of the community? The question of the enemy within is about what to do about this subgroup. Because the subgroup will cooperate internally, we can't simply isolate one free rider. Preventing the enemy within requires a different set of social constructions capable of defending the community against a dedicated subgroup. We can see the problem of the enemy within appearing, for example, when a US politician calls half of the country "real Americans," implying that the other half are not, or when the leader of one party declares their highest goal to be not the safety or well-being of the nation but rather to make their opponent a one-term president. Importantly, there is a very big difference between the statement "This president is bad for our country, and therefore my goal is to make this president a one-term president" and "This president is from the opposing party, and therefore my goal is to make this president a one-term president."

Some researchers, such as Peter Turchin, have argued that this tension across levels creates a dynamic balance between within-group and between-group effects that is a fundamental property of human history. Communities of cooperators become powerful and take over, but then, over generations, selfish members of that group (free riders and enemies within) destroy it from within, leaving it vulnerable to new groups of cooperators capable of outcompeting the uncooperative group of selfish players. As such, in addition to the free rider and enemy within problems, we need to find a way to protect ourselves from other groups of cooperators—the enemy at the gates. The question of the enemy at the gates deals with the question of how to respond to an invading army or a neighboring tribe. Sometimes dealing with the enemy at the gates is called *coalition defense* because groups that work together generally win out over groups that don't. Solving the problem of the enemy at the gates requires social tools that allow us to fight together even if we have been arguing among ourselves beforehand.

It is important to note that while sometimes the responses to each of these problems may lead to violence, violence is not necessarily the best solution. Not surprisingly, the question of violence is very complicated. Saying that violence is not necessarily the best solution does not mean it is never the right answer, only that nonviolence sometimes works. We will address the success and failure of violent and nonviolent solutions to problems as we delve deeper into these questions throughout the book. Certainly, one can execute the free rider, but once dead, they can never become a productive member of society. Alternatives (such as admonishments, fines, or banishment) provide an opportunity for redemption, and we will see that escalating punishments yield provably better moral structures than immediately jumping to severe punishments such as execution or banishment. Humans have developed social technologies allowing us to increase cooperation that do not depend on violence and instead provide opportunities for redemption, whether for a criminal to reform and become a contributing member of society, or for an enemy at the gates to turn into an ally, or for citizens to peacefully overthrow an authoritarian government through quadrennial revolution (democratic elections). Moreover, we will find that ways exist (moral systems, social technologies, aspirational goals) to make these problems unlikely to occur in the first place.

In a manner similar to what we saw in chapter 2, it is possible to change the rules of the game so people become more likely to cooperate. For example, adding in the ability to punish players who do not put money into the pot provides a means of making selfish behavior less appealing and, not surprisingly, increases cooperation.[9] Similarly, modifications in group size and group identification and the rules that define group membership can change the likelihood of cooperation. An important factor is the stability of the groups and the ability to choose groups and to move between groups. For example, if we add the ability for an individual to change groups, then people can leave selfish groups for cooperative groups, although cooperative groups will need mechanisms to prevent selfish individuals from infiltrating their cooperative group.[10] We will see the importance of issues of reputation and communication, as even just adding in discussions before deciding how much to contribute to the pot can change the amount of cooperation observed.[11] These social norms and social institutions change the rules of the game.

In the rest of this book, I will argue that this story of changing the rules through social norms and social institutions is the key to understanding the entirety of human morality. Humans have developed social institutions that interact with evolved processes to help ensure that we will be in cooperating groups and that the selfish cannot infiltrate our groups to damage them from within. I will argue that the colloquial term *morality* encompasses exactly those developed social institutions and norms and interacts with those evolved decision-making processes and learned skill sets, increasing the likelihood that we will be a cooperator in a group of cooperators. Moreover, well-designed social institutions and social norms can make this an assurance game so that increased cooperation produces improved well-being at the individual level as well.

4 RECIPROCITY

Repeated games are assurance games because defection can be punished in the next round.

Let's imagine two villages, one in the mountains that grows a summer crop and one in the valley that grows a winter crop. In the summer, the mountain people take their crop down to the valley to sell, and in the winter, the valley people take their crop up to the mountain to sell. Each time, the sellers are taking a risk—they are assuming that the buyers will treat them fairly and won't steal their food or kill them. But if the buyers were to defect, the sellers would never trade with them again. They would drive themselves from the upper-left cooperate/cooperate corner into the lower-right defect/defect corner of a repeated prisoner's dilemma game. In fact, the cheated village would likely prepare for war and fight the cheating village, putting it in a particularly negative situation. This threat helps ensure that the villages trade peacefully and thrive through reciprocity.

TIT FOR TAT

One of the most important experiments in this field comes from Robert Axelrod, first reported in a 1981 paper in collaboration with W. D. Hamilton and then in a popular book published in 1984. Axelrod had access to early computers and suggested a contest to find strategies that could work in a repeated prisoner's dilemma game. To remind the reader, in a prisoner's dilemma game there is a cooperation payout when both players cooperate, a cheater's and sucker's payout when one player defects but the other cooperates, and a mutual defection payout when both defect.

In the prisoner's dilemma, the cheater's payout is greater than the cooperation payout, the mutual defection payout is greater than the sucker's payout, and the cooperation payout is greater than the mutual defection payout. The first two conditions make it better for an individual to defect than to cooperate, while the third condition makes it better for both players to cooperate than for both to defect, which creates a conflict between the individual level (better to defect) and the group level (better to cooperate). Importantly, however, Axelrod included the important (fourth) condition that twice the cooperation payout was larger than the cheater's payout plus the sucker's payout so that his repeated game was an assurance game.

The question Axelrod wanted to test was *What strategies would be most beneficial when pitted against each other?* Researchers had been studying these games mathematically for years, but it is difficult to mathematically determine how complex sets of strategies interact. So Axelrod decided to use these new computers (it was 1980) to simulate interacting strategies.

He asked his colleagues to submit strategies. He then turned the game into an evolutionary biology population dynamics experiment. He assumed there was a population of *agents*, each of which could implement one of the strategies suggested by his colleagues. (These little miniature programs act within a simulated world. They are not real biological entities, but they do take actions, so we typically call them agents. Perhaps today we might call them *bots*.) Each agent played every other agent for two hundred rounds of this repeated prisoner's dilemma game, after which the least successful agents were removed (killed off) and the most successful ones replicated.

This is a Darwinian simulation of a species seen as a distribution over traits that shifts with environmental changes.[1] In Axelrod's simulated world, the environment consisted of the social group of other agents; the other members of one's species are a very important part of any social agent's environment. This simulation allowed Axelrod to measure whether any strategies were better than any others, where "better" meant more likely to propagate throughout the population. Moreover, he could measure whether those evolved strategies would be robust to infiltrators.

Obviously, you cannot simply have a strategy of "always cooperate" because then a strategy that "always defects" will crush you. On the other hand, "always defect" is not a viable strategy either because defect/defect pairs is an economic disaster. What Axelrod found is that a simple strategy called *tit for tat*, submitted by Anatol Rapoport, was by far the reliable winner. The tit-for-tat strategy starts by cooperating but then does whatever the other player did in the previous round. What this means is that tit for tat will cooperate with players that cooperate (including other tit-for-tat players) and will punish players that don't. Because each player interacts with lots of other players, tit for tat finds a good balance between cooperating with those who will cooperate with it and not getting suckered by those who won't. In contrast, strategies that try to cheat (defect against) it lose out because they get punished by tit for tat.

As we've seen, the repeated prisoner's dilemma is actually an assurance game, and in an assurance game, the best move is to do what the other player is going to do. If I can be sure that you are going to continue cooperating (which you can signal to me by cooperating), then I can continue cooperating with you, and we can both do well. But if you screw me over, then to hell with you. In a sense, the tit-for-tat strategy submitted by Rapoport for Axelrod's simulations took a wary risk and then reciprocated.

These wary steps toward mutual thriving (like the two villages in our parable at the start of this chapter) are something most humans are (not surprisingly) experienced with. New relationships begin with small steps that are reciprocated, and as those reciprocated steps continue to show cooperation, people become more willing to share more. These cautious steps are necessary because sharing is a risk—the more you share, the more vulnerable you are to being cheated.

We can see this wary development with any relationship, whether it be two businesses buying larger and larger orders from each other or the classic process of dating. When two people are considering each other as potential mates, they start with a process of checking each other out, often first meeting in a neutral place. People do not typically agree to marry someone they've just met. (Arranged marriages are also complex negotiations that proceed through long time courses, just not negotiations between the couple themselves. Note that an arranged marriage depends not only on developing trust between the people doing the negotiating but also developed trust between the couple and the people negotiating on their behalf.) Even casual interactions such as one-night stands include a stepping process (buying drinks, talking on a dating app) that provides opportunities to shut down the discussion before actually hooking up.

Other species that need to cooperate also show processes of developing risk assessment. For example, other primates (vervet monkeys, baboons, chimpanzees) begin with tentative steps (sitting near each other, grooming) before taking more trust-dependent steps (such as defending each other in a fight, which is known as *coalition defense*).[2] Vampire bats need to share food because they cannot go more than a few days without eating, but hunting is not always successful. Studies have found that vampire bats show this same process, beginning with tentative steps (grooming, other supports) before taking larger and larger trust-dependent steps (sharing valuable food).[3]

Subsequent experiments have looked at situations in which a decision to cooperate sometimes appears as a defection. (In our mountain/valley trade example above, maybe the caravan got waylaid by brigands on the way, or maybe the food that got sent spoiled on the way.) In this version, variants of tit for tat with punishment (if you defect, I defect twice back at you) and forgiveness (if we're defecting against each other, an occasional extended olive branch can get us back on the cooperate-cooperate track) were found to be more robust than simple tit for tat. (Notice the morality words here— *punishment, forgiveness.* We will see that intention and regret, as evidenced by speech acts such as an apology, are important factors in returning to mutual cooperation.) We will address these issues later when we come to the morality of redemption.

Tit for tat is an example of dyadic (two-person) social control. It says, "You better not cheat me because if you do, I will punish you." A community of tit-for-tat agents with lots of mutual interaction evolves to cooperate with each other and is robust to infiltrators, but tit for tat breaks down if you encounter each person only once.

SEQUENTIAL GAMES

So far, we've looked at games in which players choose their options simultaneously (the matching pennies game, the prisoner's dilemma, the assurance game, and, at the group level, the tax or public goods game), but understanding tit for tat experimentally requires sequential games. (Remember, these games are providing us with a form of "moral experiment." We can use these games both as thought experiments, asking what strategies will optimize the return of reward, and as actual experiments to measure how humans behave.) Let's start with two sequential games: the *ultimatum game* and the *dictator game*. The ultimatum and dictator games will get at questions of sharing and punishment. Then we'll introduce a third sequential game, the *trustee game*, which will allow us to examine questions of wariness and risk-taking.

The Ultimatum Game

Evidence that humans care about fairness comes from a very well-studied economic game called the ultimatum game. In the ultimatum game, two players decide how to split a pot of money (say, $20). The first player gets to decide how much to split between the two of them. And then the second player gets to decide whether to accept the deal, at which point they go home with the split, or to reject the deal, at which point they both go home with nothing.

Notice that as long as Player A gives more than $0 to Player B, it is in Player B's economic interest to accept the deal. However, in many situations this is not what humans do. In practice, most people reject anything less than about a third of the money. Knowing this, people playing Player A tend to offer between a third and half of the money to the other player.

In chapter 8 we will look at the neuroscience and the psychology of sharing in the ultimatum game. For example, it is possible to drive people to share more or to share less by changing their perception of how they gained the $20 in the first place.[4] People are more likely to reject an unfair offer if they think they are playing with an actual person (as compared to a random condition).[5] It is possible to train people to change how much they share by changing their belief in the goals of human behavior.[6] It is possible to drive people to share more or to share less by changing how they think about why they should share.[7] Different cultures tend to share different amounts on average.[8] People share more with relatives, friends, and within their community than outside it, which means that sharing in the ultimatum game depends on who you define as part of your community.[9] Finally, sharing and not and rejecting and not access different parts of our brains and can be manipulated both neurophysiologically and pharmacologically.[10]

The Dictator Game

Many variants of the ultimatum game exist, which we will discuss as they come up. But one important variant is the dictator game, in which Player A still gets to divide the money, but then the game is over. The dictator game is like the ultimatum game, but Player B doesn't have the option of rejecting the deal.

So, now, Player A shouldn't care if Player B is going to reject an unfair deal or not. In practice, people still give some money to the other player. On average, players offer less in the dictator game than in the ultimatum game, suggesting that Player B's ability to reject unfair deals is a factor in people's decisions when they are Player A. The fact that players do offer something in the dictator game means that people share even when the other player can't retaliate. Something inherent in human behavior believes in fairness.

I like to think of the dictator game as similar to tipping in a restaurant that you will never return to. The service is already over, so the amount you tip will have no effect on the waiter's behavior. There is no economic reason to tip a waiter one is never going to meet again in a restaurant that one is never going to return to. However, as with most concrete examples, this logic belies the complexity of the real world—you might find yourself back in the same restaurant someday, or you might meet the waiter again, perhaps, at a different restaurant. If you are not alone, the other people with you may see how much tip you leave. (This is a form of *signaling*.) Heck, the waiter might call out after you that you are cheap and didn't tip well, which would then affect what everyone else in the restaurant thinks of you. That last example is *gossip*, which has an effect on one's *reputation*.

Some people have argued that there is always the possibility that the other player can retaliate, so the dictator game is never truly anonymous. But this is, in some sense, the point. We will see that people construct societies where we are always playing the equivalent of a repeating game, even in situations where we encounter each other player only once.

The Trustee Game

The *trustee game* is like a repeated version of the ultimatum and dictator games. In the trustee game, one player (the trustor, Player A) has some money (say, $20). That player can give as much as they want to the other player (the trustee, Player B). When the money is transferred to the trustee, it increases (let's say it triples—imagine that the trustee can invest the money to increase it, but the trustor has to "trust" the trustee with the money). The trustee can then pay some portion of the money back to the trustor.

So if, for example, Player A gives $10 of their initial $20 to Player B, the $10 turns into $30, and Player B can give back some, none, or all of it. If Player B gives back none, then Player A has the $10 they kept, and Player B has the $30 from the tripling. If Player B gives back all of it, then Player A has $40 (the $10 they kept plus the $30 from the tripling), and Player B has nothing. If Player B gives back $10, then both players have $20 (Player A from the $10 they kept plus the $10 returned and Player B from the $20 they kept). And so on.

In the one-shot version of this game, nothing forces the trustee to pay anything back. In a sense, the trustee (Player B) is playing a dictator game with the money entrusted to them by the trustor (Player A). Nevertheless, most people do pay a reasonable amount back. Sometimes people will say they returned the money out of concern about what people might think of them, which suggests social norms of gossip and future interactions. However, people playing the trust game often say they paid some money back "because it's only fair." This opens up the very important question of *first-person social norms*—people often behave morally not because they've reasoned that it is better for them as an individual but because they have internalized a moral code. In the next chapter, we will begin to see how these internalized moral codes can develop and be useful.

In the repeated version of the game, the trustor can learn about the trustee through their behavior. The trustee wants to convince the trustor to continue putting money into the trust. In our tripling $20 example, a cooperative trustor and trustee pair could each gain $30 per cycle if the trustor puts the full twenty into the trust (which triples to $60), and the trustee returns half of it ($30), giving each player $30.

This would, of course, make the trustor more likely to put the full $20 into the trust in the next round as well. This $60 per cycle is the best the team can get if they work as a team. If, on the other hand, the trustee doesn't return any money to the trustor, then the trustor is less likely to put money into the trust in the next round—what's the point? If the trustor puts no money into the trust, the team only gets $20 per cycle (all of which stays with the trustor, who is, I guess, not very trusting).

In repeated versions of this game, partners start off wary but typically put increasing amounts into the trust and return increasingly fair portions back.[11] In particular, if the trustee finds themself in a situation in which the trustor is not providing large sums to the trust, they can "coax" the trustor back through generous returns. This is exactly what we expect to see in a dyadic development of trust. You risk small steps, and as that risk is paid back fairly, you become more and more willing to take a risk until you successfully end up in the cooperate-cooperate corner.

FROM SEQUENTIAL GAMES TO SOCIAL CONTROL

As pointed out by Lynn Stout, humans practice *passive altruism*. We don't steal from the beggar in the street. We hold the elevator door open for the stranger running to catch it. And when we find someone with their car stuck in the snow, we don't kill them and take their car; we help dig them out. In general, if someone needs help, we tend to help. Simply put, we don't actually live in a nightmare of anarchy, even without an overarching authority. We've already seen that humans tend to cooperate even in one-shot prisoner's dilemmas and in the dictator game.

In the next chapters, we will see that this occurs because we have powerful tools (such as *gossip*, *reputation*, *governance*, and *religion*) to turn many individual one-shot prisoner's dilemmas into a multiple-round assurance game, even if we play each round with a different person. Moreover, some of those tools create first-person self-control, in which one takes the cooperative road even without the danger of being punished for one's sin.

Let's imagine that Axelrod's agents playing tit for tat live in a community where they communicate with each other and where they know how the agents play against other agents in the community. First, we can include the opportunity to quit playing against someone so that someone who defects too much loses all future interactions from that encounter. Second, we can include reputation so an agent knows what proportion of cooperations or defections the other agent has chosen in the past. Finally, we can include migration so an agent can choose which agents to play each round with. Cooperating players get to cooperate for a longer duration, while players that defect get ostracized by the community ("You want to hunt rabbits, go hunt rabbits. I'm going to go find someone else to hunt deer with").

We can imagine an agent saying, "I want to be in that group—they cooperate." This raises the important question of *signaling*, whereby the members of a group (or the group itself) can say, "We're a group of cooperators. If you want to cooperate, come cooperate with us." Of course, such a group will need a mechanism to prevent defectors (the selfish) from infiltrating it (such as a community-driven method for punishing selfish players). The community will also need a mechanism to prevent groups of defectors (who only share with their subgroup and don't punish people who defect against the rest of the general community) from infiltrating it.

These mechanisms provide social control. They change the rules of the game. These social norms and social institutions can make an agent more confident that they are playing the assurance game with cooperators. But they can also be used to force an agent into a negative situation, preventing an agent from receiving the fruits of cooperation. What this means is that social control can be positive or negative. Morality is about finding those means of social control that allow agents to play their part in a cooperative assurance game.

What we need is to find a community we trust. In the rest of this book, we will see that humans are particularly good at creating just such communities. They have developed social controls that change these repeated games into a cooperative assurance game even if one interacts with each person only once.

5 ASABIYA

A sense of community helps people work together.

In a sense, the argument is that humans are evolved to work in teams, where the team rises or falls together. I find it fascinating that most human play is about teams and teamwork, whether it be children playing together or professional sports. Even most human jobs are as part of a team—think sailors on a ship, a platoon of soldiers, a corporate management team, or, like the life I live, members of a research laboratory. Humans are really good at teams.

One of the keys to a successful team is that everyone has a part to play within it. In my laboratory, I have technicians who have skill sets no one else has and who know how everything works; postbacs, graduate students, and postdocs leading individual experiments; early undergraduates doing the basic work as they learn the ropes; and senior undergraduates with their own small projects. My role is to make sure everything is on track, to keep the organization running smoothly, to guide the projects and papers, and to write the grants that keep the money flowing. (I often say that I run a small business—my lab has income and expenses. I have personnel to manage. Our product is scientific discoveries. While we don't get paid directly for each discovery, as a retail store might make a profit on each item sold, the more discoveries we make, the more we get paid. In a sense, we are like on-retainer contractors—the government pays us to keep making discoveries. As long as we're continuing to make important discoveries, the business will stay in the black.)

I have the large overview knowledge, which means that I can step into each project and help fix problems and make sure a project leader isn't missing some critical check. But my technicians are far more skilled at the laboratory techniques than I am (because many of these skills take practice and because they have better fine motor control than I do), and my project leaders know more about their projects than I do (because they are "in the trenches" looking at the data on a daily basis and because they are thinking about their one project, while I am keeping track of multiple projects). My laboratory runs smoothly because I can trust that these talented people can do their jobs as part of the team.

One can tell a similar story about lots of human teams—about a platoon of soldiers or sailors on a ship or a baseball team. Once you get to really competitive levels, you can't take a top-level catcher and make him a pitcher or vice versa. In fact, even sports that are ostensibly single-player games (tennis, NASCAR racing, Olympic athletics like gymnastics) actually depend on large teams in the background. A NASCAR racer's pit crew is an integral part of the team, and aficionados definitely know the names of the key players on those pit crews. A tennis star or an Olympic gymnast has a team of trainers and doctors keeping them healthy and practiced.

My project leaders (the postbacs, graduate students, and postdocs) have undergraduates who help them with their projects. Part of my job is to train those project leaders to become good mentors so that when they graduate and go on to found their own laboratories or to wherever their careers take them, they will become successful managers and team leaders themselves. I try very hard to make sure that the people in my laboratory do not feel like they are in competition with each other and that they know they will each achieve more individual success by helping each other. What can I do to ensure that the lab works together? How can I ensure that everyone feels they are part of the team?

ASABIYA

Peter Turchin calls this concept of social solidarity, or teamwork, *asabiya*, a term from Ibn Khaldun (1332–1406 CE), a Tunisian scholar from the Arab renaissance. The idea is that a team with asabiya will beat an equivalent team without. In fact, a team with asabiya can often beat a team of better individuals without. In any physical or economic conflict (such as between societies, cultures, or empires), working together is a huge advantage. Think of the Guppees and the Chupwalas in Salman Rushdie's masterpiece *Haroun and the Sea of Stories*, in which the Guppee army of pages debate every order beforehand but then fight together once they've come to a consensus, while the much larger and scarier Chupwala army fight as much among themselves as between and thus lose the war. The Chupwalas are so socially incohesive that they fight against their own shadows.

So what creates asabiya? The key is a sense of belonging, a sense of being part of something larger than oneself.

Beyond Kin

In early theories of teamwork, it was assumed that groups of animals worked together, sacrificing themselves for a group, because of genetic similarity. Mathematically, under the evolutionary statement that the actual atom of evolution is the gene, one can argue that a gene that (on average) sacrifices one person to save at least two others with the same gene will propagate itself into the future. Mathematically, this explains sterile workers in bees and ants and the social insects, but humans are more complicated. Worker bees are clones of each other (they receive identical gene sets from the queen), but a queen's children are not clones (they receive half their genes from the queen and half from the male that provided the sperm). A worker bee is actually more closely genetically related to her sisters than her children, making it more genetically useful to make sisters than to make children.[1] Humans don't work that way.

While kinship clearly does play a role in human behavior, it is also clearly not the answer to why humans have a society and work in teams. This has to be a function of culture, not just shared genes.

If we look at human societies, we see teams everywhere. These teams are often not genetically related. Instead, we find groups defined by culture. In the modern world, we are part of a nation (soldiers will die for their country), but we are also part of a neighborhood or city or state. We help our neighbors and pay taxes to each of these polities. Many people identify as part of a religion (tithing to the group or participating in rituals therein) or fans of a sports team. People can be members of a club, or a corporation, or a lab. The classic term *kith and kin* identifies both your relatives (kin) but also your neighbors and companions (kith).

Many people talk of their "found family"—people you are not related to but would give anything for; people who hold you up when you fall and whom you hold up when they fall. The military term *band of brothers* or *blood brothers* is about found family—about how those in extraordinary circumstances create a family-like group of individuals to support each other. One of my favorite quotes is "Home is where, when you have to go there, they have to take you in,"[2] which is fundamentally a statement about found family.

We will see later that a key to kith is your definition of your own identity. In the blockbuster movie *The Avengers*, a group of individual superheroes can only defeat the alien invaders when they finally work together, when they become something more than themselves. They do this as they learn to respect one another's individual talents and how those talents can synergize to contribute to the team's success. The key to this integration is the death of one of their own (Agent Phil Coulson), which leaves them with a martyr and a cause to fight for. The importance of sacrifice, the development of purpose, and the creation of asabiya through the definition of a new identity (here, being "an avenger" rather than an individual superhero) are classic themes seen throughout literature, in large part because they speak directly to the human condition.

Another fascinating thing about human teams is their complexity. Our teams are often fluid. Old stories often talked about "the dominance hierarchy," but while chickens and some simpler species may have strict dominance hierarchies (particularly when trapped, such as in a barnyard),[3] the real power in a primate group, such as a baboon troop or a family of vervet monkeys or a tribe of chimpanzees, is very complex. The real power in primates often depends on coalitions, family relationships, and what one might call *soft power*—the ability to get others to agree to your plans, which is often not through a physical fight.[4] Humans are even more complex—human rank is fluid and depends on context and circumstance.

We have teams within teams. My laboratory exists within a department, which exists within a university, and beyond that we are all working together within a scientific enterprise that we hope will make the world a better place (science's track record on making the world a better place is pretty damn good). But more than that, we live in an intricate web of teams. Each person in my lab is also a member of many other "teams." Some of them have spouses and children; some have parents and siblings; some are members of religious groups, and others are members of clubs; and each has a cadre of friends they spend time with outside the lab.

Beyond Dyadic Interactions

It is tempting to assume that these teams are just manifestations of shared, paired behavior. As two individuals, we can play cooperate/defect games (the prisoner's dilemma, the assurance game, and even matching pennies, ultimatum, and dictator games) with each other. If we have repeated interactions, then strategies like tit for tat will be useful to us. If you have a population of players who all interact with each other, then these strategies are stable in an evolutionary biology population dynamics sense. A group of individuals who each play tit for tat will do well and will generally outcompete interlopers, making the population stable. In more realistic, noisy environments (where cooperation sometimes appears as defection), modifications of tit for tat are necessary for stability, but the basic logic holds—a population of dyadic (two-person) interactions will appear as cooperating groups.

The problem is that this isn't what human coalitions look like. We often interact with each other only once. We do tip in restaurants we will never return to. As Lynn Stout points out, we don't steal from blind beggars on the street. My friend's kid lost her wallet at Disney World. My friend called their lost and found, and someone had turned it in, untouched, with the cash intact and with the credit card numbers not stolen. He went to the lost and found, and they gave it back to him. The person who found it never met him, and he never got a chance to thank them. As we've seen experimentally, people share more than necessary in both the ultimatum and dictator games. Human cooperation is not simply dyadic interactions playing tit for tat. Something else is going on, but what?

To answer this question, I want to go back to the tax game and the problem of the commons. Most human team interactions can be seen as a form of these two situations, whether a soldier deciding not to flee in the face of the enemy (which would likely be better for the individual but disastrous for the fight if everyone fled), or a fishing trawler captain choosing not to overfish a resource (which would be individually better but disastrous for the community if everyone overfished), or a citizen just deciding to pay their taxes (not paying taxes makes one individually richer but means there's less money for valuable infrastructure). The key is that we are not playing that simplest version of the tax game.

Modifying the Game

One of my favorite books on the question of morality is Peter Leeson's *The Invisible Hook*, in which he studies the sociology of pirates. Pirates are particularly interesting from a morality perspective because they are individuals who have forsaken the normal definitions of society and have chosen to live outside of it in what is, by definition, a criminal enterprise. However, pirates do not operate alone. They operate as a team.

First, by living on a ship, they are tied together in the basic questions of maintenance. If the ship sinks, they all drown, so it is important that everyone do their part to maintain the ship. Some of these roles are difficult and take specialized skill sets (being captain, being ship's surgeon), while others, no less necessary, are unpleasant (someone has to clean the heads and repair the sails).

Second, pirates make their living by attacking other ships. Those fights are dangerous. The first pirates over the gunwale, boarding the other ship, are more likely to be injured or killed than those following after the enemy has been wounded. As an individual, staying behind is better, but if everyone stayed behind, then the pirates would never take the other ship.

And third, there is loot to divide. How should one divide the spoils? Should the strongest pirate take the most? Should it be divided evenly? Should someone with a special skill set get more of the reward? What about people who were particularly brave in the fight or severely injured? Should they be rewarded differently? What about people who stayed behind or showed cowardice in the fight?

The key to creating a group dynamic that will outcompete other groups in the assurance game is to ensure that cheaters cannot succeed within the group. As we saw in chapter 3, within the group the selfish outcompete the cooperators, while groups of cooperators outcompete groups of selfish players. So what can one do to prevent players from being selfish? What can pirates do to make sure no one stays behind when they attack a ship?

THIRD-PARTY PUNISHMENT

This brings us to the concept of *third-party punishment*. Third-party punishment is when one person punishes another for cheating a third—you played unfairly to others in our group, so I'm going to spend my resources to punish you and make you regret your actions.

This Is Not Just Coalition Defense

It is important to differentiate third-party punishment from *coalition defense*. Coalition defense occurs when a group bands together when attacked. Coalition defense occurs when you have a group (say, with two agents A and B), and a third player (C) attacks a member of the coalition (C attacks A), and the other member comes to their defense (B attacks C). Third-party punishment occurs when the three members are all part of the same group (A, B, and C), and B punishes C for being unfair to A within the group.

Coalition defense is certainly a staple of human behavior. Peter Turchin has argued that coalition defense is the key to the construction of empires—tribes fight among themselves until they are faced with a common enemy. The Romans and Etruscans were at each other's throats until the Gauls poured out of the Alps with Hannibal in 216 BCE and destroyed the Roman army at Cannae. (While everyone talks of Hannibal bringing the elephants over the mountains, it wasn't really the elephants that defined the battle; it was the army of Gauls that Hannibal's envoys had convinced to join them.[5]) But once the Romans were able to regroup, they began to work together with the Etruscans and the rest of the Italian peninsula to defeat Hannibal, until they discovered they had a new (Roman) empire. Coalition defense consistently pushes people to work together. A group that is attacked defends itself as a group. (This is why many people have suggested that the best thing for world peace would be an alien invasion, including the famous fabrication of one in the graphic novel and movie *Watchmen*.[6])

There are lots of examples of coalition defense in nonhuman animals. Some of these coalitions are defined by simple relationships, like kinship—for example, parents defending children (*Don't mess with mama bear's cub!*) and the many examples, particularly among primates, of siblings defending each other. As pointed out by numerous primatologists studying primates in the field, primates such as baboons and vervet monkeys are very aware of their kinship relationships and will coalition-defend along those relationships (*Don't mess with my kid brother!*). Extensive examples can also be found of nonsibling coalitions within primate tribes and groups (*Don't mess with my friend!*). These are held together by mutual grooming and other friendship-related behaviors.[7]

However, the evidence for third-party punishment in nonhuman animals is limited.[8] Some of the most interesting examples of these limits are Jane Goodall's observations of the inability of the Gombe tribe of chimpanzees to handle the sociopaths Passion and Pom. Passion and Pom were a mother-daughter pair (a kinship-related coalition) who would find a mother chimpanzee with an infant, harass the mother until she was separated from the infant, and then kill and eat the infant. The males of the tribe would interfere if they saw this happening but did not treat Passion or Pom differently in other situations, and after the deed was done, they would sometimes partake in the cannibalism.[9]

Humans do not (generally) put up with these kinds of crimes. While people sometimes get away with crimes against children, no one defends those crimes as good or agrees that getting away with those crimes is acceptable. Humans build structures and institutions to prevent and punish such crimes.

Crime and Punishment

The classic example of third-party punishment is that of punishing criminals, which is a central part of human culture. Anthropological studies have found that bands, tribes, and groups of humans in general have laws and norms to enforce punishment by the community. Certainly, one of the purposes of government is to punish criminals in order to increase the cost of taking actions that the community does not want taken.

Barry Reilly and his colleagues studied the viability of robbing banks as an economic job.[10] The average bank robbery takes in only about £12,000 (it was a UK study), so one would have to successfully rob two to three banks a year to make an average wage (the median wage in the UK in 2019 was about £30,000). Even given a low probability of getting caught (say, one robbery in five, or 20 percent), a bank robber is more likely than not to be in jail within a couple of years. This makes bank robbing a particularly poor job prospect. The key, of course, is the chance of getting caught and going to jail; if bank robbing were not illegal, then it would be a profitable job.

So we, as a community, say that robbing banks is unacceptable and provide third-party punishment (going to jail) as the price of that choice.

In general, the solution to the tax game problem (the selfish do better within a cooperating group) is to create social norms and institutions that produce third-party punishment. Because the institution is doing the punishing, every individual pays only a little to construct that institution. Because of social norms, individuals know they won't be allowed to get away with cheating. We will see in chapter 6 that there are a number of advantages to making the punishment come from the group rather than the individual, but for now, we can assume the enforcement institution is part of the central tax, which cooperators are willing to pay to keep the selfish limited.

Ostracism

The ultimate punishment is to remove someone from the community. This can certainly be accomplished by killing the offender, but that is a very permanent solution to what is hopefully only a temporary problem. The more common solution is to ostracize them, to separate them from the group and mark them as "nonpersons" for purposes of the community (think of being fired from a job or being released from a sports team. In earlier times, someone might be run out of town).

Ostracism removes the offender from their support system. It says, "We are no longer going to play the assurance game with you" ("No more deer for you—good luck hunting rabbits"). Given that the world is a dangerous place and given how hard it is as a human to survive alone, one expects that ostracism was (and is) a particularly devastating punishment. The hope is, however, that the ostracized offender will see the error in their ways and return to the fold, begging for forgiveness.

Generally, punishment processes are constructed to be impermanent—they provide an opportunity for the individual to redeem themselves, usually, through actions and requests for forgiveness. For example, the parole system in the US depends not just on an individual having served the minimum time, but also on evidence that the individual has changed, has a plan for a return to society, and, essentially, wants to be redeemed.[11]

If we look, for example, at successful groups with third-party punishment, what we find is an escalating process, with small steps (proportional responses) for first offenses and escalating punishments for future offenses. The idea is that the first offense provides an opportunity for the offender to express regret, to ask for forgiveness, and to show that they can change. For example, the offender might be given a warning, then a small punishment, and then an escalating punishment, as offenses are repeated. If the warning is enough, then the offender can become a valued member of the society. Importantly, as offenses build up, it is necessary for the offender to work harder and harder to regain the trust of the society. An important step in this process is the human emotion of shame, which is the feeling that one has offended the culture.

It is interesting to compare the constructs of *shame* and *guilt*. Of course, concepts of guilt and shame vary across cultures, which makes them part of that culture's *moral toolbox*. Furthermore, we will see that these concepts interact with our decision systems to affect our behavior. Guilt is related to the emotion of regret. Guilt is the recognition that one has committed an offense (a crime) and entails taking responsibility for one's actions, presumably with some remorse for those actions. In contrast, shame is the recognition that society has identified one's failure. Guilt is a recognition of the offense, but shame is the recognition that one is part of a team and has let the team down. Guilt and shame interact in interesting ways with respect to intention. One is unlikely to feel guilt for an accident or a failure that is not one's fault, but one can certainly feel shame for an accident or a failure to accomplish a necessary goal as a result of one's inadequacy. I will address the complexity of accidents and intention in the discussion of "Moral Praise and Moral Blame" below.

Guilt is also interesting because it can occur even in situations in which one has not been caught, which raises the possibility of *first-person self-control*, whereby a person refuses to cheat even when given the opportunity to do so. Social structures that create guilt can help push people away from defecting even when they won't get caught. We will explore the usefulness of first-person self-control, even for the person expressing it, in chapter 6.

Nevertheless, what this third-party punishment process means is that team members are watching out for each other. We are not simply pairs of individuals playing dyadic (two-person, tit-for-tat) games.

It is important to remember that people are members of multiple teams that interact. In Francis Ford Coppola's *The Godfather: Part II*, Frank

Pentangeli's sacrifice for the mob family is rewarded through a promise of permanent support for his wife and kids. Pentangeli is scheduled to testify to the government commission about the details and mechanisms of the Corleone mafia enterprise, which would presumably be enough to allow the government to dismantle it, but he recants his testimony and commits suicide instead. His death by suicide is a sacrifice for his other team—his family. Notice that this requires a level of trust in his mafia compatriots because Pentangeli will not be there to ensure compliance with the deal. This is technically a prisoner's dilemma that Pentangeli has to play as an assurance game. As a reward for his sacrifice, Michael Corleone is going to spend resources helping Pentangeli's family. But, economically, it would cost him less to cheat Pentangeli. Michael Corleone could tell Pentangeli that he will provide for his family and then not. But Pentangeli is not concerned about Michael Corleone keeping his promise. Why not?

Reputation and Signaling

The key turns out to be language and reputation. If you cheat me, even in a single-round prisoner's dilemma, I can let it be known that you are unreliable. The next person to interact with you will know that you cheated me. When we carry this out to the larger set of interactions, we turn multiple two-person prisoner's dilemma games into a public goods tax game.

If we have a community of agents playing repeated cooperate-defect games with knowledge about how agents have played in previous rounds, then an agent who defects loses even more because no one wants to play with them anymore. (This is a form of ostracism.) This knowledge about how agents have played is *reputation* and depends on both signals from the agent in question (trying to show how they are going to play) and information from others about how that agent has played in the past (*gossip*).

The notion of reputation brings us to the discussion of signaling. One of the points that Leeson makes is that the classic pirate flag (the skull and crossbones) was a signal to enemy ships that the pirates would let the prey live and return home safely, presumably with enough food, and with an intact ship, but only if they didn't resist. If they did resist, then the pirates would kill everyone.

Notice that this is another case of modifying the payout matrix. Without the flag, the payout matrix looks like the following:

Table 5.1				
The payout matrix with no signal from the pirates				
		Prey		
		Surrender	Fight (lose)	Fight (win)
Pirates	**No flag**	**Prey** loses all goods, maybe die too **Pirates** gain all goods, no pirates die	**Prey** loses all goods, maybe die too **Pirates** gain all goods, some pirates may die	**Prey** keeps goods, some may die **Pirates** lose goods, some may die

From the merchant ship's (the prey's) perspective, fighting is better than not fighting. If they lose, they are in no worse shape than if they surrender, and if they fight, they just might win. On the other hand, even if the merchant crew chooses to fight and then loses (i.e., the pirates win), the pirates risk getting injured and may lose some of their personnel. It is definitely better for the pirates to get the merchant ship to make the Surrender choice. In other words, they need to make it worthwhile for the merchant ship to choose Surrender rather than Fight and Lose. So the pirates signal to the prey that there is a different payout matrix:

Table 5.2				
The payout matrix under the skull and crossbones				
		Prey		
		Surrender	Fight (lose)	Fight (win)
Pirates	**Skull-and-crossbones flag**	**Prey** loses all goods, *all prey survive* **Pirates** gain all goods, no pirates die	**Prey** loses all goods, *all prey die* **Pirates** gain all goods, some pirates may die	**Prey** keeps goods, some may die **Pirates** lose goods, some may die

Suddenly, surrendering becomes a viable option (literally).

Notice that the pirate-prey interaction of the skull-and-crossbones flag in table 5.2 is one of reputation. The flag signals to the prey that they are part of an ongoing repeated interaction. Even if the pirates encounter this specific prey ship only once, when the prey return back to shore, they will tell everyone that the pirates honored the flag—they took the goods but let everyone go. A pirate's reputation is key, both in fear and honor. This leads to one of the running jokes in the movie *The Princess Bride*, that the Dread Pirate Roberts (not just the Pirate Roberts, but the Dread Pirate Roberts) is actually retired and living comfortably on his takings. As Westley says when he asks if Inigo Montoya has "considered piracy" as a new job, "It's the name that counts. Not the person, you know." (It's all about the role to play. We'll talk about roles more later.)

Gossip

The key to reputation is that there must be a way to communicate that reputation across the community. The point of reputation is that instead of Person A having to be cheated by Person B to know not to trust them, Person A can use the fact that Person B cheated Person C as a reason not to trust Person B, even before interacting with them. So how does Person A find out about Person B? Gossip.

Experimental studies playing the public goods tax game have consistently found that groups in which players are allowed to talk to each other share more than groups that cannot talk to each other.[12] This talk within the group about who is cooperating and who is not is fundamentally a question of sharing information about the reputation of individual players and is a form of gossip. Reputation is an even more effective defection-prevention mechanism than third-party punishment because third-party punishment occurs after the crime, but reputation prevents one from making the mistake of trying to cooperate with someone who is going to defect against you before they even try. The combination of the two is a particularly effective social control system.

One of the most intriguing things about human language is that most of it is gossip.[13] What is gossip? It's telling stories about reputation—who is cheating who (or cheating on who—not surprisingly, sex and reproduction are key human evolutionary concerns). Gossip has been a staple of human dynamics probably since the beginning of language. From ancient stories to modern magazines, from villagers talking over the fire to friends dishing on each other in pubs, to women passing around lists of which men not to trust at work, gossip is fundamentally information about reputation.

The key to gossip is that it is a centralized pool of information. Person A does not have to talk to Person C to get the scoop on Person B. Person C can tell Person D who can tell Person E who can tell . . . and the gossip goes around until everyone knows that Person B is untrustworthy. Of course, gossip can be used as a weapon itself—Iago can drive Othello to violence by lying about his wife. But, eventually, a metalevel of gossip can occur, and one can learn not to trust gossip from the "mean girls" lying about each other. Furthermore, in normal human cultures gossip provides an opportunity to morally condemn or praise an individual, and we are generally not tolerant of hypocrites who claim their own moral stand but tear down others.[14]

Gossip, however, does depend on seeing the deception and being able to report the consequences. Importantly, this means that if no one knows about one's sin, there is no one to report back and no gossip to be passed around. This is why pirates would leave one person alive from ships that fought back—to go home and tell everyone the deal. Fight and die or surrender to the skull and crossbones and live.

MORAL PRAISE AND MORAL BLAME

There is, of course, a huge literature on what makes something moral or not and the factors that should be incorporated into that decision—whether generalized rules are intrinsically moral or not, whether the motivation matters or merely the outcome, and whether an individual can define their own morality or whether morality must be agreed upon by a group. This philosophical literature is not built on a scientific tradition and turns out not to be very useful as a foundation on which to build our study of morality. However, we can deconstruct this literature to examine what it tells us about the psychology of human views on morality. Instead of asking what these philosophers are telling us about how to think about morality, we can use our observations of their logic to determine how humans decide what makes something deserving of moral praise or moral blame.

In general, moral blame and punishment accomplish three goals. First, there is recompense, through which we make it less worthwhile for an individual to defect (play selfishly) in our community game. Thus, if we find that someone has not paid their taxes, we can make them pay both the taxes and the cost we spent in catching their defection. Second, there is specific deterrence, in which an individual is deterred from reoffending. Escalating punishments make it possible to catch early defections and hopefully convince the offender to mend their ways before they become truly dangerous, while still providing a strong deterrent for serious defections. Third, there is general deterrence, in which other individuals are warned against committing a crime after having seen the consequences. General deterrence in order to potentially prevent defections before they occur is, of course, why blame and punishments are often made public within a community.

Importantly, deterrence assumes intention and the hypothesis that punishments will have consequences. Thus, we should differentiate between harm caused by accidents and harm caused intentionally. Furthermore, we should differentiate between accidents caused by negligence and accidents caused by a lack of knowledge or bad luck. For example, we should probably treat a car crash caused by cut brake lines differently than one caused by failed brakes worn-out through normal use. Furthermore, we should probably treat a car crash caused by worn-out brakes you were warned about but didn't fix differently from a car crash caused by an invisible patch of black ice (transparent ice on a black road that creates an invisible but very slick surface).

Similarly, understanding the psychology of motivation is an important factor in determining how to handle punishment and deterrence. In chapter 7, we will see that our actions arise from many different decision systems that learn in different ways and thus respond differently to different social structures. Understanding how these decision systems interact with our (social) environment is important to creating good social structures. As one example, viewing drug addiction as a disease to be treated or as a consequence of a lack of sociological opportunities allows us to create social structures that produce better outcomes than those derived from viewing addiction as a crime to be punished. (We will discuss the consequences of changing moral views on addiction in chapter 13.)

Recent studies have found a fascinating difference between what people typically praise for moral success (moral praise) and what people typically criticize for moral failure (moral blame). Moral blame tends to be about ways one should change one's behavior, taking into account causality, intention, and the magnitude of the outcome, while moral praise tends to depend more on indications of character, taking into account motivation more than the actual consequences ("at least they tried"). Ineptitude can lead to moral blame, particularly in terms of the consequences of carelessness, but accidental morally good actions tend not to be praised. Similarly, morally suspect reasons tend to preclude praise. This makes blame generally depend on the consequences of the action, while praise often depends more on the intention of doing good.

This is exactly what one would expect from our view on social control and asabiya. Moral blame is a statement to the community that a behavior requires third-party punishment, while moral praise is a statement to the community that a person is of good moral character and should be allowed into the club.

Someone's Watching

If no one sees you, then there's no reputation or gossip to be found. This suggests that people are more likely to play fairly if they are being watched. But people do tend to help, and they do tend not to cheat even when no one is watching and even when they are not going to get blamed for cheating or praised for not cheating.

How do we know this? How can we test to see if people are playing fairly when we can't directly observe them? This would seem to create a paradox. Even if an experimenter tells a subject they're not being watched, the subject might not believe them, especially if there's a one-way window set into the side of the experimental room or a camera in the corner.

What we can do is have our subjects do something random, like roll a die, flip a coin, or play pachinko (a game in which balls fall through a series of pegs and should distribute across the bottom in a "normal" or bell-curve distribution), and not show anyone the answer. The key to the experiment is to have different outcomes pay different amounts. (Say when the subject reports the coin flip, heads pays out at 10¢, but tails pays out at $1.) We ask the subjects to tell us how their coin flip came out (which no one else saw). Over enough subjects, the reports should be fifty-fifty—half the subjects should report heads, half tails. To really control for things, we can tell half of the subjects that tails pays out $1 (and heads 10¢) and the other half of the subjects that heads pays out $1 (and tails 10¢). (This is called *counterbalancing* in psychology experiments.) If people are more likely to report their $1-paying choice than their 10¢-paying choice, then we know some people are lying, but we don't know who. Since everyone plays only once, you can't tell whether an individual is lying or not. Rare events do happen. But, because each player's game is independent, we know how the samples should distribute (one out of six on the six-sided die, fifty-fifty on the coin, and as a normal bell curve in the pachinko game). So you can look at the distribution of what everyone reported and compare it to the expected distribution.

Experiments examining these and other opportunities for cheating do find that when no one is watching, people are more likely to cheat, but they also consistently find that most people are honest even when no one is watching. Moreover, on games in which people can cheat a little or a lot (with a larger payout depending on the rarity of an event), most people who do cheat only cheat a little. There are two important points here. First, most people continue to play honestly even when given the opportunity to cheat. Second, people do care about being watched.[15]

The theoretical structures we've built over the last chapters predict this second point—observation should matter. Humans are evolved to respond instinctively to being watched. For example, one of my favorite experiments is the classic social science experiment by Melissa Bateson and her colleagues, who measured the willingness to pay for coffee and milk in an academic office coffee service.[16] The coffee and milk were in the break room, and people could come by whenever they wanted and take some but were supposed to put money into the kitty to cover the costs. No one was there to physically sell the coffee; it was all on the honor system. The experimenters could measure how much money was in the kitty and how much coffee and milk got used, so while they couldn't measure who had played fairly and who had cheated, they could measure the amount of cheating. Then they did something wonderful. They put a sticker above the price list. The sticker was either of flowers or of watching eyes. When the sticker was a pair of eyes watching, fewer people cheated.

We will see in chapter 15 that one role religion plays is to ensure someone is watching at all times. ("God is everywhere." "Your ancestors are watching over you." "Santa knows if you've been bad or good.") Researchers have suggested that this is one of the main roles played by early religions—to provide that constant surveillance and ensure that people behave as if they're being watched. However, this is only one step toward an explanation, as even atheists are generally unlikely to cheat. In fact, there is no evidence that people who are atheists are any less moral than people who are theists.[17]

Many people continue to play fairly even when they won't get caught. To answer why that happens, we need to break down the concept of social control carefully.

6 SOCIAL CONTROL

Humans have cultural (and evolutionary) tools capable of helping a community of humans ensure that they are living as a community of cooperators.

In chapter 5 we found that individual and community practices, the responses made to each other's choices, changed the way members of a given community behaved. These social norms and structures are a form of social control. Studies of actual communities (fishing communities in Chile or Alaska, lobstermen in Maine, nineteenth-century whalers, small indigenous hunter-gatherer bands, or ranchers in Northern California) have found that community control is a complex interaction of cultural norms that encourage cooperation, enforced through one-on-one interactions (termed *second-party dyadic punishment*, such as tit for tat, paying retribution so as to even up a disparity), and *third-party social control*, with hierarchical legal systems stepping in when opportunities for community social control are lacking. Key to all of these successes are aspirational goals that drive an individual's desire to cooperate (*first-person social control*) and differentiate communities.[1]

We can think of this social control as existing at three levels. We can start with *second-party dyadic punishment*, in which one person punishes another for being unfair. Dyadic social control depends on repeated interactions. But in situations in which you only interact with another person once (or rarely), dyadic social control is inadequate. Instead, we can develop *third-party social control*, whereby a community gathers together to punish offenders and to create laws that enforce coordination. Third-party social control depends on centralized (and accepted) social institutions, as well as metanorms that the third-party solutions will be enforced. Third-party social control depends on community, and the differences in community lead us to *first-person self-control,* in which an individual internally decides to cooperate even given the opportunity to cheat. First-person self-control depends on cultural norms, aspirational goals, and assumptions of shared group identity.

Let's examine each of these in turn.

DYADIC (PERSON-TO-PERSON) SOCIAL CONTROL

Two of the most interesting studies of how social norms change behavior are James Acheson's *The Lobster Gangs of Maine,* which details sociological interactions on the Maine coast, and Robert Ellickson's *Order without Law,* which details sociological interactions among ranchers in Northern California.

Both of these studies found cultural norms of reciprocity within the community, whereby a favor should be returned by an equivalent favor, and more importantly, a trespass (whether it be accidental or intentional) should be returned by an apology and restoration (which allows for redemption). Ellickson notes that the cultural norms encourage approximate equivalence, such as one rancher doing extra fence maintenance and the other watching the first rancher's house while he was gone for three months out of the year. (Ellickson's study is about cattle ranchers in the 1970s, and the fences we are talking about were miles long, had to be checked and maintained at regular intervals, and could take days to check, riding mile after mile on horseback.)

Ellickson found that the equivalent nature of the reciprocity was important, even if the trade might be in different currencies. Reciprocity that is not equivalent could be seen as making statements about hierarchies. Interestingly, both too small and too large returns can be seen as problematic—too small indicating that the apologizing party does not see the wounded party as worthy of respect and too large indicating that the apologizing party is of higher rank and providing largess to lesser parties.

Ellickson notes the anti-monetary nature of the favors, requiring that favors be made "in-kind." Returning the favor in-kind helps maintain a sense of equality between parties because paying recompense in money was seen, culturally, as placing the payer above the recipient sociologically. Ellickson notes that returning favors in-kind forces the interaction to be more personal. It is important to recognize, however, that these interactions are culturally bound. There are cultures where monetary recompense is recognized as valid and acceptable. For example, in medieval Norse and Icelandic cultures (as described in the sagas), atonements for violent crimes were made through monetary fines to the surviving parties, although those restitutions were not always successful at stopping the ensuing feuds. Similar monetary compensations can be seen in extant cultures as well.[2] Economically, giving a gift certificate or providing money should, theoretically, allow a closer titration to the actual cost of the favor, but it also separates the people involved in the transaction. Within the community studied by Ellickson, monetary compensation was seen as a business deal rather than a community construction between friends. Moreover, monetary compensation assumes that one can put a simple financial value on the favor, which may be difficult.

As an example of the danger of providing a simple price for a given trespass, one can look at the famous case of the day care group that tried to solve the problem of parents picking kids up late by charging them for it.[3] Picking kids up late from day care is a selfish act that puts one's own time as more valuable than that of the caregivers (who only agree to watch your kids until the day care closes at the end of the day). Because the day care providers cared about the kids, they did not want to just close the day care and leave the kids out on the street to wait for the late parents. (And let's acknowledge that the day care providers would probably have been punished by the community had they abandoned the kids at the end of the day.) In order to punish parents who arrived late, the day care providers decided to charge them a fee. But then some parents accepted the fee as a reasonable price for arriving late, so they just arrived late and paid the fee. In fact, the day care found that more parents arrived late once the fee was instituted than when the "punishment" was complaints from the day care providers. One could argue that the fee simply wasn't high enough, but a higher fee might have led the parents to argue that the fees were exorbitant, as many cultures have rules against price gouging. Price gouging is a classic example of selfish anti-community behavior and would have shifted the sin from the late-arriving parents to the day care providers.

As a solution, Ellickson notes that the members of the cattle-ranching community did favors for each other that were "close enough" in value, as determined through cultural interactions. By not trying to put an accurate monetary number on the favor, one can trade favors that are quite different and noncomparable, such as watching someone's house for three months and maintaining a multi-mile-long fence.

One-on-one (dyadic) control of selfishism is fundamentally about changing the economic games that we have been discussing. If someone cheats you in a repeated game, playing tit-for-tat forces the next trial to be an "I'm not going to cooperate with you" trial, which pushes them into a bad situation. Yes, it also pushes you into a bad situation, but this can be seen as a form of mutually assured destruction. It says, "I'm willing to forego the opportunity to cooperate with you because I don't trust you." So if Player B has defected against Player A on Trial 1, then a tit-for-tat Player A defects on Trial 2, limiting Player B's opportunities. In Trial 2, in the example prisoner's dilemma game we laid out in chapters 2 and 3, Player B cannot get the $10 win at all. Player B can only get a $0 win, giving Player A $10, or Player B can defect, giving both players $1.

Table 6.1						
Dyadic punishment in the repeating prisoner's dilemma						
		Player A		$\rightarrow$	**Player A**	
		Cooperate (C)	**Defect (D)**		**Defect (D)**	
Player B	**C**	**A** and **B** each get $6	**A** gets $10 **B** gets nothing		**A** gets $10 **B** gets nothing	
	D	**A** gets nothing **B** gets $10	**A** and **B** each get $1		**A** and **B** each get $1	

If Player B is smart, they would like to find a way to get back in Player A's good graces. We can think of this as a negative in Player B's accounts with Player A that need to be corrected. So what does one do when one makes a mistake and sins against another?

We all know what one needs to do. Young children are taught this in every human culture. You admit your mistake, apologize, and provide some sort of restoration. Ellickson calls this "even-up" and identifies multiple examples of this in the rancher community, whether something as simple as helping mend the fence that cattle broke, loaning the use of equipment, or providing the gift of a valuable commodity. Interestingly, Ellickson also provides examples of one party taking the even-up from another, such as when one rancher "borrowed" another's backhoe without asking in return for cattle trespass on his land. In these situations, the resolution can be as simple as saying, "Yes, now we're even" and not responding to the tit-for-tat defection. Of course, this "borrowing" could also be seen as "theft," leading to spiraling mistrust, which is a large part of the danger of dyadic controls on interactions.

Importantly, if the offending player (Player B in our scenario) believes that the response (by Player A) is out of proportion to the sin, then the balance has shifted, and now Player B is the offended party. This can create a feud, whereby two parties alternate back and forth with escalating destruction. The problem, of course, lies in the emotional response to the sin—one is full of anger and wants revenge. We will see in chapter 9 how emotional responses can lead to disproportionality and how one needs to use social tools that shift the responses to other decision systems that can more accurately respond proportionally and can recognize that the proportional response allows the possibility of redemption.

It is tricky to find the proportional response in tit-for-tat punishments. One option is to enact community norms that can be worked out for repeated crimes that occur often. For example, Ellickson found that it was acceptable to respond to a cattle trespass situation (where cattle break a fence and end up on another person's land) by driving the cattle to an inconvenient location but not to kill the cattle. Killing the trespassing cattle would be seen as a disproportionate response and would require a return response that could turn into a long-term feud. Similarly, lobstermen fishing in Maine are very aware of the danger of escalation, so they prefer to damage the traps found in one's territory rather than scuttle a boat or burn a dock. However, when crimes are extreme or tinged with emotion, it can be very difficult to return with a proportional response that does not escalate the situation.

Literature is full of examples of attempted proportional responses that fall apart because of strong emotions that drive an internal goal for escalation and revenge.[4] Importantly, a proportional response does not preclude future scenarios, but it gives the enemy the option to say, "We've evened-up, let's cooperate again." It provides a path back to redemption if the other side will take it.

Tit-for-tat individual social control assumes there will be repeated interactions to force one into the assurance game. But often we interact with a given party just once and do not expect to interact with them again. Dyadic (individual, one-on-one, tit-for-tat punishment) processes are inadequate for those interactions. If we live in a community, however, we can create third-party social institutions that shift these interactions into a cooperative regime.

THIRD-PARTY SOCIAL CONTROL

A solution to both of these problems (emotional responses produce escalation, and a lack of repeated interaction makes tit-for-tat punishment irrelevant) is third-party social control. In studies of the tax or public goods game, repeated games in which it is possible to punish selfish players (who do not put money into the central pot) lead to greater contributions and thus greater individual and group success.[5] However, in these scenarios the punishing player is paying a cost to punish the selfish player, which makes applying punishment a second-order public good and subject to its own form of free riding—maybe I'll let you punish the selfish player so that I don't have to spend my own resources to punish them. One option is to have metanorms, whereby players who don't punish are punished themselves, but this just moves the free riding to the next level. A much better solution is pool punishment, whereby the group punishes the selfish player. In fact, when given options between games in which individuals punish selfish players or games in which groups punish selfish players, people prefer the group-punishing games.[6] In these group-control games, the restoration is driven by the community itself, not by the individual. This is why, for example, in the law, criminal actions are identified as *State v. Perpetrator* rather than as *Victim v. Perpetrator*. It is the community (the state) that is punishing the crime.

Many communities believe the apology itself should be to the community because by creating intracommunity strife, the perpetrator has damaged the community. Presumably, the victim will be satisfied with the punishment meted out. Some communities, such as the Quakers, would often require transgressors to stand and apologize publicly to the community, while other communities, such as the Catholic Church, demand that the transgressors apologize to God privately. Nevertheless, note that these codes identify proportional responses for each transgression. By enshrining the proportionality in the code itself, we do not need to depend on an individual to recognize or identify the proportionality, which, as we've noted above, is often difficult when emotions are involved.

The general goal, assuming that one believes it possible, is to reform the transgressor and keep them in the community. A reformed transgressor can be good for the community if they become a cooperating player in the assurance game. (Remember that these games are not zero-sum. The more cooperating players we have in our community, the better our community does.) The anthropologist Chris Boehm describes a situation in which a hunter cheated his community during a group hunt in the jungles of central Africa. This tribe of foragers hunted for food by placing large nets in the jungle and driving large game animals (gazelles, antelope) into them. The perpetrator placed his net in front of the others and thus captured an unfair amount of meat for his family. They punished him by taking the excess meat and also the stored food the family had stashed away. He apologized profusely to the community and was allowed to continue participating in future hunts. Note the importance of *redemption*—the offender was a capable hunter who would be an advantage to the community if he stopped defecting against it and started cooperating.[7]

This community control is termed *third-party punishment*, and it forms a set of laws or community norms whether they are written down or not. For example, Ellickson reviews nineteenth-century norms among whalers that determined who was able to harvest what under what conditions, which changed depending on the difficulty in capturing a whale. In the early days, whales were likely to be chased down by a single ship, and the (unwritten) norms were that the first ship to harpoon the whale had claim to it. However, as whaling shifted to include whales more likely to require being run to exhaustion and thus more likely to escape one whaler only to be captured by another, the (unwritten) norms changed to share whales between the ships that had both been necessary for the successful hunt. Similarly, Elinor Ostrom's extensive work on common pool resources identifies how different communities have different norms that relate to the specific properties of the common pool resources. And we are all familiar with the way that people who transgress the norms of a community can be shunned or ostracized—such as unwillingness to let someone participate in the church social.

Laws and Regulations: The Problem of Strangers

Dyadic social control requires repeated interactions between individuals. Third-party social control requires repeated interactions within a community, even if you only meet any individual once. But what is to be done in situations in which the community is very large and repeat interactions are rare? For example, what should be done when a tourist from the city hits a rancher's cow that has wandered onto the road, totaling the car and killing the cow? Here, we have two people who are unlikely to have extensive future repeat interactions either with each other or with others in their individual communities. The solution, as noted by Ellickson, is a legal structure that both parties have agreed to (which belies the title of his book, *Order without Law*). Ellickson found that fault and recompense depended on structures put in place by various governmental entities (at the county, state, and federal levels) and in this case hinged on the presence or absence of fencing built by the rancher as well as the attention paid by the driver.

Moral codes often talk about the importance of being kind to strangers and not only cooperating within your community but also being a generally "moral person" and helping out those strangers. But, of course, experimental studies of humans find that people are less willing to cooperate with people outside of their group.[8] This is called *parochial altruism*—the idea that we are altruistic (cooperators) within our parochial group and competitive (selfish) beyond it. Many authors have argued that parochial altruism is the descriptive behavior of humanity. And, in fact, we will see in chapter 10 that human neurophysiology has evolved to identify groups and to determine how much more one should share within one's group than beyond it.

Moreover, a naive view of the group assurance games described in chapter 3 could lead us to a similar conclusion—that one should only share within the group. However, this ignores the fact that these assurance games are not zero-sum and that the more people we bring into our community the more people there are to cooperate with us and the better off we will be as individuals within that society.

The goal of these games is not to create enemies beyond the group but to create cooperators within the group. Creating enemies beyond the group is only a tool in the moral arsenal. It is not the goal. As we saw in chapter 2, like Westley and Inigo Montoya we are trying to figure out how to become friends. It is undeniable that human societies have gotten larger over the millennia, with more and more people being brought into larger and larger cooperating groups capable of reaping more of those non-zero-sum benefits.

As communities get larger and larger, we need social constructions that enable us to build those cooperative communities, even with strangers. Importantly, those social constructions will need to take into account the human neurophysiology that determines whether we see someone as trustworthy or not, whether we regard that stranger as a friend or an enemy, and whether and to what extent we are willing to share our resources with them.

Social and physical technologies allow us to create larger and larger cooperative groups. Dyadic punishment works in groups of repeated interactions, reputation and gossip provide for small communities, and laws and regulations enable larger groups. Laws and regulations are agreed-upon social norms that transcend individual interactions. However, it is not only the social constructions that have enabled larger group sizes but also the physical technological developments, such as an increasingly visceral news media (oral histories and stories, which developed into printed media and then to radio and television and now to social media) that can provide for increased understanding and empathy across vast distances. We will see in chapter 10 that personal stories can increase empathy and trust. Seeing someone's story in a movie can make us more willing to empathize with them and include them in our concept of our community. Furthermore, scientific developments can also change our willingness to see strangers as community, such as the recognition that race and gender are fluid sociological constructs that do not align with the older physical hypotheses that once suggested castes and limiting roles.

Of course, not all social norms are positive. These tools can also be used to create enemies and push individuals apart. Importantly, these more negative social constructions are generally about whose good counts, about who is part of the protected class, and about who gets to be an agent in the assurance game. I will address these issues of the morality of group norms in the last chapters of the book when we compare the implications of policy decisions on groups and individuals. In short, we will find that tolerant communities that include more members (and more diverse members) are often more powerful than intolerant ones. Moreover, more diverse and tolerant groups are often better even for individuals who would be in the dominant class in the intolerant groups. Importantly, these social norms interact with our psychology; shifting our moral codes from negative ones to positive ones will require the development of tools that take that psychology into account. A good example of that interaction can be seen in the question of dissent and protest, which will lead us to the concept of aspirational goals.

Dissent and protest fit directly into the logic being built in this book. Dissent and protest are moral tools that can change societies. As with the other moral tools, they can be used for good or ill. We will return to this question when we address the problem of moral relativism and argue that the economic games we have been looking at provide us a way out of the problem of moral relativism (chapter 18), although we can already state that it is useful when examining dissent and protests to ask whether the changes they are advocating for increase or decrease the cooperative community. Like the individual demanding reciprocity in table 6.1, protesters demand societal changes before they are willing to come back to participate in society. Dissenters and protesters are not free riders; they are acting members of a community insisting that it change how it plays the assurance game. Similarly, whistleblowers insist that these social rules be applied fairly and that cheating individuals not be allowed to get away with cheating. Protesters, dissenters, and whistleblowers are using moral tools to change societies and to change how we play the assurance game within that society. Martin Luther King Jr. wasn't trying to destroy America; he was trying to save it.

Revolution and the Question of Violence

For many years, philosophers thought the only cultural tool that could maintain order within a large group was a centralized authority, but studies of actual human order have found that humans in general do not submit to centralized authority and will often step up to prevent authoritarians from taking control of a group. In fact, authoritarians who take control of large civilizations do not take control alone but rather depend on alliances and a tight-knit community within the larger community. While these authoritarians may be the face of leadership and may hold severe power in that their whim may be law, they depend on others within the community being unwilling to hold them to account, often because of largess passed down to that select subcommunity. Authoritarian regimes need to be seen not as a hierarchical leadership but rather as a community conflict, in which a tight-knit internal community takes control of the social norms of a larger community and uses those norms against it.

Whenever discussions of these sorts of issues come up, the question of violence appears. As pointed out by Boehm in his book *Moral Origins*, communities need social structures to handle both the cheat (the free rider) and the bully who uses violence to gain advantage. Studies have found that communities in general prefer to start with nonviolent responses that escalate to violence when the offending party does not respond to the nonviolent controls. To quote the civil rights leader Malcolm X, "Non-violence is fine, as long as it works."[9]

In a fascinating study, Erica Chenowith and Maria Stephan found that nonviolent revolutions were more likely to succeed than violent ones.[10] (Violent movements succeeded only about one-quarter of the time, but nonviolent movements succeeded more than half the time.) Moreover, this study did not include prescheduled revolutions, such as those seen in practicing democracies. Elections are a form of social technology. A social construct of a mandated national election changes the rules of the game; it allows a leader of an opposing party with a different perspective on governing to replace an authoritarian government through accepted processes. It makes possible revolution without violence, as long as all parties respect the outcome. Respect for that democratic election is a moral code. We will see later (chapter 19) how we can create moral structures, such as elections, to encourage competition without conflict.

In large part the increased success of nonviolent movements has been due to their ability to recruit much larger portions of the population into the movement. Chenowith and Stephan found that large nonviolent movements could topple regimes from social pressure alone because of the interactions with other cultures and other societies. A key to the success of these movements is that they reveal information about oppression to the rest of the world, which opens up third-party responses that can put pressure on bullies. See, for example, the successes of the nonviolent protests led by Mahatma Gandhi in India in the 1930s and 1940s, the American civil rights movements of the 1950s and 1960s, and the anti-apartheid movements across the world of the 1980s. The fascinating thing about these movements is how they brilliantly used the news media as a tool to bring on third-party punishment and social control. In the American civil rights movement, the sight of fire hoses, dogs, and billy clubs attacking unarmed protesters galvanized a more powerful national community capable of coming in and enforcing the social equality enshrined as an aspirational goal within the community. Similar effects are now being seen through the ubiquity of cell phones and social media, allowing the community to witness the violence that has been perpetrated in our communities for decades (centuries). Importantly, these scenes have angered members of unoppressed communities, who have responded by saying "That is not who we are" and "That is not who we will be." (Similar issues are at work in the success of the recent Black Lives Matter and #MeToo movements.)

It is important not to underestimate the moral force of aspirational statements in these situations. The civil rights movement depended on the aspirational documents that underlie the US, including the statements that "all men are created equal" and have "unalienable rights" to "life, liberty, and the pursuit of happiness." Of course, as is well known, the US has never achieved those aspirations—not among its founders, nor when Martin Luther King Jr. and the civil rights movement argued for equality, nor now when we are only beginning to acknowledge the existence of *systemic racism*. Nevertheless, the existence of these statements as aspirations provides a means for the downtrodden to appeal to third-party goals, to defend dyadic retribution, and to encourage change within individuals, leading to first-person self-control. These aspirational documents help define the cultural norms that individuals see within themselves. Importantly, those third-party structures interact with individual first-party self-norms, making it possible for someone to appeal to another person's private view of themself to change their mind about a moral issue. (Interesting examples of this are Abraham Lincoln's journey to the Thirteenth Amendment and Lyndon Johnson's journey to the Civil Rights and Voting Rights Acts.[11])

Nonviolence works because it calls on third-party punishment within a shared community. It depends on assumptions of shared aspirational values. When the others with physical power do not feel a part of that shared community or do not share those aspirational values, then nonviolence is unlikely to work, leaving violent conflict as the only response.

Of course, violent conflict usually leaves both communities devastated and has (in the economic language we are using in this book) high *transactional costs*—additional costs that are not part of either component of the interaction but are due to the interaction itself. Thus, in our mountain/valley trading example from chapter 5, there are real costs of growing the mountain crops and growing the valley crops, but there are also transaction costs of carrying the crops up or down the mountain. Because it is presumably cheaper to grow mountain crops in the mountains and valley crops in the valley, trading allows the two communities to grow more crops with less effort as long as the transaction costs are not too expensive. Not only does a marching army destroy the community it is marching across (e.g., Europe in World War II, Sherman's march through Georgia, the Russian and American invasions of Afghanistan, or the state of Iraq and the Middle East after the US interventions of the 1990s and 2000s), but war is expensive. The Cold War bankrupted the Soviet Union and the current war on terror is bankrupting the US. If we can find a way to enforce social control without violence, we can reduce the transaction costs and increase our community success.

Furthermore, while it is possible to find a way back to redemption and cooperation after violence, that journey can be long and difficult. For example, Germany, Japan, and Vietnam are all now US trading partners, but this achievement took decades after those wars ended. It is much easier to find a path to redemption and cooperation if the change is driven nonviolently within the community rather than violently between communities.

The success of nonviolence depends on aspirational documents and the structure of community. The existence of these norms that we are discussing makes it possible for a victim who is outmatched physically, fiscally, or otherwise to defend themselves by effectively appealing to the larger community at hand.

FIRST-PERSON SELF-CONTROL

This brings us to the aspect of social control that is the most difficult to establish within the economic language we've been using so far. However, it is also the most important when we start to talk about morality. The idea of first-person self-control is that one has a moral code, a set of goals and rules that guide behavior even if no one is watching and there will be no dyadic or third-party punishment. To quote Michael in the TV show *The Good Place*, it's that little voice in your head saying, "Come on now, you know that this is wrong." Moral blandishments are often phrased in terms of first-person self-control, stating that one should behave appropriately (cooperate with the community, be kind to strangers) whether anyone is watching or not.

One important driver of first-person self-control is that it can be difficult to know whether one is going to get caught. Thus, first-person self-control can provide a form of precommitment, a way around the temptation of thinking that one might get away with being selfish. Furthermore, if one is only going to defect in situations where one really can't get caught, then one will have to be extremely vigilant about determining the likelihood of getting caught. This takes mental effort. As we will see in our discussion of the neuroscience of decision-making in chapter 7, people have limited cognitive resources to spend on these kinds of social and environmental monitoring. At some level, if you are in a group of cooperators, it may be easier and less costly to just cooperate as well. It may be that the small gains obtained from defecting and being selfish just might not be worth the cost of that vigilance.

It is true that people are more likely to cooperate if they are subject to punishment for being selfish. However, many people do continue to cooperate even without the threat of punishment. As we saw, people contribute more in the ultimatum game than in the dictator game, but people do contribute to their partner in the dictator game. People share even when they don't have to. Why?

First, notice that there are transaction costs to both dyadic and third-party punishment. When a player must punish another player for defecting against them, it costs both players. Tit-for-tat says, "I will defect this time because you defected last time." Remember that the defect/defect corner is worse than the cooperate/cooperate corner. (That's the whole point of the assurance game.) Similarly, third-party punishment is a tax on the community. It makes the community pay for that punishment. If everyone in the community just cooperated in the first place, it would be better for everybody. While this description of reducing transaction costs may sound like a form of *group selection*, it is not. Instead, it is about individuals constructing social structures that improve the assurance game. Remember, in the assurance game it is better to cooperate than to defect if your compatriots are also going to cooperate. In the assurance game, cooperating is better for both the group and the individual.

This means that a community of people who each exercise first-person self-control does better than a community of people who only cooperate to avoid the threat of punishment. What we will see is that much of what we call morality (including philosophical, legal, religious, and governmental structures) is about convincing people that first-person self-control is in their own best interest.

While, economically, one could argue that someone expressing first-person self-control is a sucker, foregoing opportunities, the truth is that in a world where individuals have the opportunity to migrate between communities, communities composed of individuals expressing first-person self-control would be attractive communities to join. However, we would need a way of ensuring that the community really does consist of first-person cooperators who refuse to cheat even in conditions where they could. That suggests that communities will need markers (signs) that people can express to show they are willing to be part of that community.

The key is that our social structures do not exist in a vacuum; they interact with our decision-making systems. Humans are made up of multiple decision-making systems, each of which learns differently, processes information about the world differently, and comes up with different answers in different situations. In the next section (chapters 7–11), we will explore how these decision-making systems work and interact with moral questions. Most importantly, we will find that how the question is set up changes what information is available to the decision-making systems and thus changes which decision-making system drives behavior. This alters the actions (moral or immoral) the person is going to take. This means that we can create social structures (norms, institutions) that can actually drive first-person self-control and create successful communities where people don't cheat even when given the opportunity.

II HOW WE CHOOSE

7 THE NEUROSCIENCE OF CHOICE

Our actions are the consequences of multiple, interacting decision-making processes.

Over the last hundred years, scientists studying the neuroscience and psychology of human behavior have discovered much about the mechanisms that underlie our decision-making. In particular, over the last fifty years it has become clear that we are not unitary decision-makers. Instead, our brains contain multiple decision-making processes that are evolved to be optimal for different situations.

In my book *The Mind within the Brain*, I define a decision as "taking an action"—that is, any time you take an action, you have made a decision.[1] (Of course, not acting when one could would also be a decision.) This is important because earlier definitions only looked at decisions in which one consciously deliberated between options, but it turns out that a lot of our actions are not due to deliberation, including many that we definitely want to call decisions (such as whether to run away or to stand and fight, or whether to swing at a pitch in baseball or cricket). A number of moral actions are also not due to deliberation (such as whether to dive into a freezing river to save a drowning stranger).

Defining decision as action-selection changes the question from *How do we decide?* to *What is the information process that leads to taking the action?* In general, a decision depends on representations of the past (memory), the present (perception), and the future (the things we would like to achieve and the things we would like to avoid). Importantly, these representations are not always explicit.

One of the key discoveries over the last fifty years in computer science is that how one represents information changes what one can do with it. To understand this discovery, let's imagine a bookshelf with the books sorted by author (as many are). This is particularly useful if you've discovered a new author you like and want to find more books by that author, or more books in a series. It's less useful if you want to use the books for other purposes, such as to shim an unbalanced table or as a stand to raise your computer monitor to a comfortable level to play a computer game. For these purposes, you'd like the books sorted by size. Importantly, both options (sorting by author, sorting by size) are fine, but which option is best depends on what you need the books for. We can think of these two different orderings of the books in your library as representations. You need different information to find a book in your library under each ordering system.

I can measure how you've sorted your books by asking you to do a specialized task designed to separate these different representations: *What books come just before and just after the one you just read?* If you've sorted them by author, the adjacent books are likely to be by the same author or by authors with last names that start with the same letter; if you've sorted them by size, they are likely similar sizes. Moreover, the books should be in appropriate order: for the author-sorting system, the book before yours, the one you've just read, and the one after yours should be in alphabetical order. Similarly, for the size-ordering system, the three books should be in appropriate size order (largest to smallest or smallest to largest, depending on the ordering). In neuropsychology, we call this a *probe trial* because it probes the representation.

Let's imagine that I teach you a simple game. There are symbols on a computer screen. If an orange circle appears, type the A key, and you'll get a reward, but if a blue square appears, type the L key, and you'll get a reward. (A and L are on opposite sides of a QWERTY keyboard, so they make good keys to type.) You get really good at this game. And then I mix it up: I include blue circles and orange squares, so now what do you do? Is it the color or the shape that matters? If you've represented the difference by color, you'll make one set of choices (L for blue circles, A for orange squares); if you've represented the difference by shape, you'll make a different set (A for blue circles, L for orange squares). If you've represented the combination, you might freeze up as you try to figure out what to do with these new conditions. How you represent the data not only changes how easily you can find the answer but also what answer you give.

It turns out that mammalian brains (including yours!) have multiple representations about the past (memory), present (perception), and future (goals, motivations, desires, fears), and one will take different actions in different situations depending on which representation gets used and how.

So what does all this have to do with *morality*? Any moral action you take is an action and thus, by our definition above, a decision. So if we want to understand how humans decide to take moral actions, we have to understand how they take actions in general. But it also means that if we can change how you've sorted your books, we can "nudge" your behavior;[2] we can make it easier or harder to take a given action. For example, if we wanted to make you more likely to encounter diverse authors, we might not sort the books by authors' last names. But in order to nudge your behavior appropriately, we need to know how humans represent and use information in their action-selection systems.

MULTIPLE DECISION SYSTEMS

There are three key action-selection systems in the mammalian brain that we need to talk about. I am going to call them the *deliberative* or *planning system*, the *procedural system*, and the *instinctual system*. Technically, reflexes are also actions, and thus, by our definitions, reflexes are also a decision-making system. But reflexes don't play much of a role in moral actions, so I am not going to talk about them in-depth in this chapter. (I suppose a reflex action could create accidental harm that one might need to apologize for, but in that situation, I think the moral question is how to apologize and make recompense for that action. I don't think the moral question is whether the action should have been taken in the first place.) Importantly, complex automated responses to stimuli (like running from a lion) are not "reflexes"—they use a different neural circuit than actual reflexes like putting your arms out when you slip. Those more complex automated responses (running from a lion, fighting with a friend, freezing in fear, and falling in love) are part of the instinctual system. These three decision systems (four, including reflexes) go by many names in many literatures. In particular, much of the decision-making literature calls the instinctual system the "Pavlovian" decision-making system. I refer the reader to my earlier book *The Mind within the Brain* for a thorough description of the relationship between this computational taxonomy of three systems (four, including reflexes) and that used in other literatures.

Each of these systems represents information about the past (memory), present (perception), and future (outcomes to achieve, outcomes to avoid) differently, uses a different computational algorithm on these representations to select the action that is going to be taken, and is thus optimized for different situations. Importantly, they each use different neural circuits within the mammalian brain, which means we can use neurophysiological techniques to measure which neural circuits are active in a decision process and determine which systems are involved in different decisions.

Planning (Deliberation)

Deliberative systems use an algorithm of searching through imagined outcomes. We really do run simulations in our head[3]—like a war game, we try out possibilities—*What if I take this action? What would happen? Is that what I want?* The first key to deliberation is that the process steps through the possibilities serially. When deliberating, you imagine a single possibility at a time and then evaluate it. So, for example, if you're trying to decide whether to go skiing in Colorado or to visit Chichen Itza in Mexico for your family vacation, you imagine each option separately. You imagine taking your family to the ski slopes, renting skis, riding the ski lift, and feeling the cold and the wind and the rush of speed as you ski down the mountain. Separately, you imagine Chichen Itza with its impressive stone buildings and its hot, humid jungle, recognizing similarities and differences between its ancient cities and our own modern ones. Importantly, deliberation is a serial process. You don't imagine some quantum mix of skiing down the slope of the pyramid at Chichen Itza—you imagine them separately. This makes deliberation a slow and effortful process.

The other key to deliberation is that you evaluate these simulations by running them through the same parts of your brain that evaluate present rewards and punishments. That means you are evaluating that imagined world in light of your current needs and desires (not in terms of imagined needs and desires). This allows deliberation to consider those current needs and desires: go get food when you are hungry; go get water when you are thirsty. However, this creates interesting consequences when your needs now do not necessarily comport with your needs in the future. For example, I have noticed that I buy more food at the grocery store when I am hungry, most likely because when I imagine eating the food, I evaluate it using my current emotional state (being hungry) and think it more valuable than I would have if I weren't actually hungry now.

Deliberation is slow and serial and includes a logical determination of the action sequence that gets you to an outcome, an explicit representation of that outcome, and an evaluation of that explicit outcome in light of your current needs and desires.

Procedural Action Selection

Procedural systems learn to recognize situations quickly and then release a well-practiced action sequence (think sports). For example, a quarterback in American football has to decide whether to throw a pass or not. A tennis player needs to return a serve. A volleyball player has to get under the ball to set it or spike it. In general, these actions have to be executed much too quickly to be deliberative. Instead, we need a system that can recognize a situation (even a complex situation) quickly and release an appropriate action in response.

The procedural system learns to recognize situations through a categorization process. Neural systems are very good at recognizing patterns. They store information about expected patterns in the connections between neurons so that when a pattern is put into the network, the network is able to recognize and identify that pattern.[4] The mammalian brain contains a hierarchy of these recognition components, with each component tuned to recognize more and more abstract and generalized patterns. This is similar to the now-popular *deep-learning* artificial intelligence process, which was modeled on mammalian visual recognition systems.[5]

The specifics of this process are not critical at this time. What matters is that in procedural learning one part of your brain learns to recognize the pattern in the world that allows you to define and categorize a situation while another part learns an action chain to release at that moment. Let's take hitting a baseball as an example. Two components need to be learned: how to recognize the pitch, both when the ball will cross the plate and where it will be when it does, and how to swing the bat. Swinging the bat is an action chain that requires a specific stance, a step forward, a turn of the hips, and a specific muscle motion of the arms, all of which have to be executed smoothly and proficiently. We learn this motion through lots of practice, starting from awkward movements that would never move a baseball even if you did connect with it. Over time, with enough practice, your swing can become smooth and powerful. Recognizing the pitch is also a learned process, which depends on experience recognizing the spin on the ball, a pitcher's hand motions, and the timing of the pitcher's release. With lots of practice, it becomes possible to react in a few hundred milliseconds and to swing in order to connect a precise bat location (a few inches of "sweet spot" moving at seventy-five miles per hour) to a precise ball location (a couple of inches of appropriate contact point moving at one hundred miles per hour) and hit a home run.

In general, procedural actions have evolved to be quick to execute, even if they are slow to learn. But being slow to learn, they are also hard to unlearn, so we call them *habits*. Importantly, there are good habits and bad habits. At the moment of executing the habit, there is little conscious control (except in the sense of contextual preparation, like the knowledge that swinging a baseball bat in a baseball game and in a bar fight are very different action chains that should only be undertaken in very different circumstances), but the decision to practice to learn to recognize the situation and to execute the action sequence requires a great deal of conscious control—it takes thousands of hours to learn to hit a baseball well.[6] It takes effort and preparation to be ready for that less than a half second of decision time.

Instinctual (Pavlovian) Action Selection

Instinctual (Pavlovian) systems are about learning to release an evolutionarily prewired action at the right time. You can take a thousand tries to learn to hit a baseball, but running from a lion—you've got to get that right the first time. Every species has a set of actions that are "prewired" into each individual in that species. This limited repertoire of available actions was learned over evolutionary time (Run from lions!), but these actions can be linked up to new situations. So, for example, you can learn that when a lion is stalking you, the lion's motion produces a rustle in the grass, so you can learn that when you hear a rustle in the grass, you should feel fear and know to run away from that rustle in the grass—giving you a jump on fleeing and making it more likely that you will get away.

We call these *Pavlovian* systems because they are how Pavlov's famous dogs learned to salivate to cues that predicted food. The dogs learned that a certain set of cues (situations) would occur before they were fed. So when those cues occurred, they salivated, preparing to eat. This is how your cat learns to come running when it hears you open a can (of, it hopes, cat food), or your dog starts barking when the garage door opens.

The limited Pavlovian (instinctual) action repertoire is the set of species-important actions that have evolved to make the individual do better in the world. These tend to be key survival actions (approach food, flee from predators) and key reproductive actions (the mating dance of two people at a bar trying to decide whether to go home with each other that night, a mother smiling at a baby the first time she holds the child in her arms). What is interesting about these actions is that, in humans, they contain a surprising set of social behaviors. So, from a positive perspective, laughing with your friends is a Pavlovian action. But so is the fear of a stranger walking toward you in a dark alley, punishing someone for being unfair to you, diving into a freezing river to save a child, and deciding to donate your kidney to a stranger.[7]

Because those social interactions are the key to morality (as we saw earlier in this book), moral interactions are often fundamentally Pavlovian in their execution. We will see below that a lot of the moral actions that we talk about are Pavlovian, including reactions to unfairness, reactions to being watched, and reactions to other kinds of social and group-behavior cues. Singing together or dancing as a group produces emotional (Pavlovian) responses, which is why choirs sing in church, people dance at festivals, and armies march together during training.

The evidence that these three systems are fundamentally different—processing information in different ways, responding differently in different situations, and depending on different neural circuits—is overwhelming. Whether these three systems are the only action-selection systems is, however, controversial. For example, some researchers have argued that there is an additional system based on remembering and using past events as precedent to make a decision, but it is unclear if this is just another way to look at deliberation.[8] Similarly, if the whole decision process lies in recognizing the situation, does it need to go through a separate procedural system? Could the cortical state representation instead be directly connected to the motor structures? There is mounting evidence that such situation-recognition systems directly drive behavior in primates (including humans).[9]

Mounting evidence points to humans, at least, having a fourth system based on the reasoned following of explicit rules that are linguistically coded (delineated in language). A rule-based system would take in the information about the situation at hand and then act on it. (Such a system would depend not only on the current situation one finds oneself in but also on one's definition of oneself—"I am the kind of person who follows this rule.") Rule-based systems can change quickly, taking into account not only the perceptual stimuli within the environment but also contextual information and even episodic information. Neurophysiologically, rule-based systems can arise from pattern-completion processes. Rules based on more and more complex situations (stimuli, context, complex episodes) could come from these pattern-completion processes applied to more and more abstract representations.[10]

DUAL-SYSTEM THEORIES

The idea that we are not unitary decision-makers has a very long history and goes back to at least the ancient Greeks, who argued for Dionysian (hot) and Apollonian (cold) decision systems.[11] For example, Plato made an analogy to a charioteer trying to control two horses. This is immediately recognizable as Freud's ego trying to control the id (hot-blooded Dionysian decisions) and the superego (cold-blooded, logical Apollonian decisions). Technically, I suppose that Plato's and Freud's theories are three-process systems because they suggest that we have an executive (the driver of the chariot, the ego) who has to mediate between the two sides of oneself, but in these theories, there are only two action-selection systems.

Other versions of the dual-process theory are based on self-control, in which a logical self tries to control an illogical, emotional self. These other versions really are dual-process theories in which the controlling executive is also the logical side. The classic description is the elegant horse-and-rider analogy, in which the person is a rider trying to control a horse. The horse is the wild, "animal" side, while the rider is the logical "human." This dual-process theory goes back to Augustine of Hippo (Saint Augustine), who argued that the rational (City of God) has to control the animal (City of Man). Reading Augustine makes clear his obsession with control and lack of control—he spends pages worrying about the consequences of having stolen a pear from someone else's orchard and notes how in his youth he asked God to "Make me chaste. But not today."

Behavioral economics often talks of "animal spirits," to use the term first put forward by John Maynard Keynes; John Coates talks of the difference between the wild wolf and the domesticated dog.[12] Probably the best-known dual-system theory in recent times is Daniel Kahneman's System I and System II. In Kahneman's formulation, System II is capable of planning, logic, and reasoning. That obviously maps to our deliberative system. However, Kahneman then lumps everything else in as System I.

The problem with these dual-process theories is that the instinctual and procedural systems are as different from each other as they are from the deliberative system. In fact, Kahneman and the other dual-process theorists actually include a number of other computational processes in these heuristics (i.e., System I, non-System II), such as *priming*, a process in which decisions are guided toward cues with properties similar to those one has been thinking about or has recently experienced. Priming arises because of the computational process through which our brains recognize patterns, which makes similar and associated patterns easier to retrieve and recognize than different and novel patterns.[13] These dual-process theories are like trying to understand genetics by differentiating mammals from other life forms, ignoring the fact that fungi are as different from plants as either of them are from animals and that fungi are more different from plants than reptiles and mammals are from each other. Instead, the data support a multiple decision-systems theory that is much larger and broader than these dual-process theories. There are at least three separate decision systems and multiple support systems that allow things like priming (which depends on the specifics of how perception happens). The important point for our new science of morality is that all of these separate computations matter for the consequences of our moral tools.[14]

In the popular literature, this recognition that there are both fast reactive and slow deliberative systems has led to the concept of "making decisions in the blink of an eye," but it is extremely important to recognize that quick decisions are only useful in certain situations. In some situations, Pavlovian action-selection systems are more suitable than deliberative. (It may be easier to dive into the icy river when all you are thinking about is saving the kid and not about how cold you are going to feel.) And situations exist in which procedural action-selection systems have been trained to take the right actions after gaining sufficient expertise. But these are specific situations aligned to those specific decision systems. A medical student shouldn't make a diagnosis in the blink of an eye. A rookie firefighter shouldn't trust their instincts. The reason a seasoned doctor can often rapidly make a better diagnosis or the ground commander of a firefighting team can quickly recognize a collapsing building is because they have spent decades training their pattern-recognition systems. The opposite of being analytical in these situations is not that one is intuitive but rather that one is being glib, which can be very dangerous. This is also why it is not enough to divide decision-making into two processes (a slow, cognitive System II and everything else lumped into System I); one needs to understand the underlying information processing performed by the decision-making systems to determine when and where each is best suited to be used.[15]

One of my main complaints about these horse-and-rider models is that they suggest you are the rider but not the horse. Jonathan Haidt says it is so hard to control the horse that maybe we should think of it as an elephant rather than a horse. David Eagleman calls the other components "zombie subroutines" and suggests that the self is a conscious controller trying to tame these demons. Michael Gazziniga suggests that the self is trying to rationalize behaviors it does not control. Robert Kurzban describes us as a press secretary for a company explaining away behaviors that are not our fault.[16]

However, as can be seen from our earlier description, you are the interaction of all of these decision systems. When a sports star is "in the zone," they describe themselves as being in their body, not out of it. (They are describing the procedural system in a good flow.) When your heart starts pounding while you are walking home through a dark alley in an unfamiliar city, that's you feeling fear, not someone else. (That's the instinctual system on high alert.) In fact, it is extremely disturbing to find oneself dissociated from one's instinctual drives, which is sometimes seen in response to repeated trauma.[17] Dissociation is very negative because (as we will see later in the book) the instinctual system is a key to our membership in a community, and dissociation makes it hard to find one's place in that community. (Not impossible—it is possible to get better—but dissociation is a harsh solution to a difficult problem.) You are all of these systems, and in the horse-and-rider analogy, most readers identify more with the rider than the horse.

To be fair, at the end of Haidt's book he admits that you are both the horse (elephant) and the rider. And Eagleman is not arguing that these zombie processes are not part of your brain. In general, these arguments are more about the interpretation of who we call the self rather than the underlying neurophysiology of brain, mind, and behavior. Neuroscience and psychology has discovered, very clearly, that we have multiple algorithms making decisions that can produce inconsistencies in our desires and decisions. This inconsistency provides a key to understanding how morality interacts with these decision-making systems.

So to summarize where we are in our understanding of neuropsychology: There are at least three decision systems (deliberative, procedural, instinctual) that access different neural systems and depend on different computations because they represent past (memory), present (perception), and future (outcomes to achieve, outcomes to avoid) in different ways. They may even select different actions in a given situation. All three of these systems play roles in the kind of actions that relate to our interactions with each other. Moral and immoral actions are, by definition, actions, and thus which system is driving the decision will have important effects on our morality. These decision-making systems and the various support structures needed (like how we recognize situations by recognizing patterns) have underlying computations that change how potential moral codes affect us.

8 MORAL AND IMMORAL ACTIONS

How you ask someone to do something changes the actions taken because of how the question interacts with our decision-making processes.

Moral and immoral actions are by definition actions and, thus, decisions. As we saw in chapter 7, our decision-making systems process information in specific ways. That is, the way we make decisions depends on an interaction between the environment we are experiencing (including our social environment), what we have learned (and how we have learned), and which decision-making system is driving behavior at each moment. Importantly, changing the way we ask the question can change the answers we give. Let's take a look at some specific moral questions and see how changing the question can change the outcomes.

SACRIFICE FOR THE GREATER GOOD

The most famous morality experiment in psychology was done by Stanley Milgram in the 1960s, in which he asked ordinary working people to administer an electric shock to another person even to the point where the other might die. The story is usually told in terms of obedience—ordinary people were surprisingly willing to obey the authority of an experimenter who told them to do horrible things. Milgram decided to pursue this experiment because he was trying to get at the question of what was, at the time, known as the "good German." After World War II, when the Allies discovered the genocide of the Holocaust, they put the concentration and death camp guards on trial for "crimes against humanity," and the usual defense was that the guard was "just following orders."

We will leave for another time the whole issue of the rules of war, what is permissible and what is not, the moral sacrifices that soldiers make in times of war, and the difficult edge of morality that aligns in war. Here we will assume that the guards knew they were killing unarmed innocents. I think this is a fair assumption as no guards tried the defense that the murdered millions were enemy combatants who deserved to die. (Compare the classic play *The Man in the Glass Booth* or the Aamin Marritza speeches in *Star Trek: Deep Space Nine*'s devastating episode "Duet.") Instead, the guards generally fell back on the idea that they were just "following orders."

When Milgram started his experiments, the war criminal Adolf Eichmann had just been convicted. In his trial he seemed shockingly human. People were disturbed by the idea that these Nazi monsters were very ordinary. So Milgram set out to test whether ordinary people would commit egregious sins.

The experiment itself was elegant and simple. The subject was brought into a room with the experimenter and another person whom the subject believed to be a fellow subject. Subjects were paid $4 for their participation and, very importantly, were told they would receive the $4 no matter what happened in the experiment itself, even if they quit partway through ($4 in 1961 is worth approximately $30 in 2020). Unbeknownst to the real subject, the other person was actually an actor. The two subjects (the subject and the actor) each drew a slip of paper from a bowl to provide some randomness. Supposedly, one slip said "teacher," and the other said "learner."

This step is underestimated in most descriptions of the Milgram experiment. It means the subject likely thought they could have been the learner just as easily as the teacher. This puts us in a Rawlsian situation of equality before birth. The philosopher John Rawls has argued that a moral system requires that decisions about what the rewards and punishments will be for each position in a society must be made before one knows one's position within that society.[1] Several of the subjects in Milgram's book said that one reason they were willing to go through with the shocks is that the "learner" also volunteered for the experiment. This makes the Milgram experiment actually different from the Nazi death camp guards, who almost certainly never imagined themselves on the other side of the fence.

In actual fact, however, both slips said "teacher," and the actor lied about what their slip said. So the subject was to be the teacher in the experiment. The actor would be the "learner."

The subject was brought into a room containing a complicated box bearing lots of switches and dials. There were thirty buttons on the box, below which were labels that ranged from "slight shock" to "very strong shock" to "extreme intensity shock" and then to "danger severe shock." The last three buttons were simply marked "x x x," and the line connecting the descriptions ended in a vertical stop just before the "x x x," suggesting one should not go beyond that point. The subject was hooked up to the machine, the lowest button pressed, and a light shock delivered to the subject. Theoretically, this would convince the subject that the machine was real and would deliver shocks as labeled.

The subject was then unhooked from the machine, and the learner was taken to another room, hooked up to the machine, and strapped into a chair. The two subjects were separated so that they could not see each other and could not communicate (remember that one of the subjects is really an actor). The real subject (as "teacher") was then told to read pairs of words to the learner. The learner was supposed to respond with the "correct" paired word. If the learner got the pair wrong, the teacher was supposed to press the button to shock the learner. With each wrong answer, the teacher was supposed to press the next button in the sequence so that the shock increased by fifteen volts each time. The truth, however, was that the machine was fake, the other subject was an actor, and no shocks were delivered.

Initial estimates were that only a few subjects (maybe three out of a hundred) would shock the subject dangerously, but in the initial experiments, as many as 65 percent of the subjects shocked their compatriots all the way through to the last "x." Milgram was so surprised by this that he then tested a second condition in which the actor responded to the shocks to convince the subjects they really were hurting someone. Milgram found that this didn't change the willingness of the subjects to deliver the shocks. Some subjects proceeded through the entire sequence, even in the second condition when the learner (actor) screamed, banged on the wall, complained they had a heart condition, and went silent.

Milgram's study is usually described in terms of obedience, and Milgram's book describing his experiments is titled *Obedience to Authority*. But it's not clear to me that this is the right way to think about what happened. First, of course, 65 percent going through to the final condition means that 35 percent did not and rejected it. Second, many subjects questioned the experiment, inquired in concern about the safety and health of the compatriot, and showed tension, anxiety, conflict, and stress over the experiment. They were not simply obeying orders.

Milgram had a set of responses prepared for these questions. In response to specific questions about the health of the learner, Milgram reassured the subject of the safety of the experiment: "While the shocks may be painful, there is no permanent tissue damage." This was not a lie; the experiment was completely safe for the actor. If the subject suggested quitting, Milgram said, "Please continue" or "Please go on" followed by "The experiment requires that you continue." Additional questions led to the response "It is absolutely essential that you continue" and, finally, "You have no other choice, you must go on." These requests are not military orders. They are polite requests not to ruin the experiment. Notice that the second prompt draws its demand from the experiment, not the experimenter. The third prompt talks about the essential nature of continuing, presumably due to the importance of the experiment. Only the last statement has any relationship to an actual order.

Milgram included several transcripts from the subjects in his book. Many subjects argued back, responding to the prods. These subjects responded to the explicit order of the fourth prod with anger, not obedience. An industrial engineer responded to "You have no other choice" with "I *do* have a choice. (*Incredulous and indignant:*) Why don't I have a choice? I came here on my own free will. I thought I could help in a research project. But if I have to hurt somebody to do that, or if I was in his place, too, I wouldn't stay there. I can't continue. I'm very sorry. I think I've gone too far already, probably" (emphasis and stage directions in the original). A professor at a divinity school responded to the fourth prod of "You have no other choice, sir, you must go on" with "If this were Russia, maybe, but not in America."

We don't have quantitative data on which prompts led to quitting from Milgram, but in a replication study done by Jerry Burger in 2009, 64 percent of the subjects continued in response to the first prod ("Please continue"), 46 percent after the second ("The experiment requires that you continue"), and only 10 percent after the third ("It is absolutely essential that you continue"). No one continued after the fourth ("You have no choice, you must go on"). The more the prod sounded like an order, the more people reacted negatively to it, which is consistent with psychological theories of how people react to threats to the ability to make their own choices.[2]

Of course, there were some major differences between Burger's replication study and Milgram's original study. For example, the replication study only went to 150 volts (which Milgram had labeled as "strong shock") and found slightly fewer people willing to reach this level (70 percent in the 2009 study; 82.5 percent in the original Milgram study). There were other differences as well, including differences in the demographics of the subjects, in the awareness of the news of the day, and in the trust in governmental authority figures. Milgram's study was before the Vietnam War, Nixon, Iran/Contra, and the weapons of mass destruction excuses for the invasion of Iraq, while Burger's study was after all of them.

Nevertheless, subsequent experiments Milgram ran suggested that the authority of the experimenter did matter, but we have to be careful about what we mean by "authority" here. People were more willing to follow the shocks through to the end if the experimenter wore a lab coat than if not, if the experiment were held at a university rather than a storefront, and if the experimenter spoke with confidence than if not. But these differences are less about authority in the sense of "having power over" and more about authority in the sense of "being a respected member of the scientific community." The subjects were more willing to shock the learner if they were confident that this was a real and valid scientific experiment.

Milgram did report that during debriefings afterward, many of the subjects talked about the importance of the experiment, not the experimenter, as the reason for continuing. There is more going on here than "obedience"—people were willing to play out the game for reasons beyond simply obeying orders.

We have to place Milgram in his time. The first experiments were done in 1961. This was the height of the Cold War, which Americans believed they were fighting with the Soviets. The battlefield was not soldiers fighting and dying on the beaches of Normandy or in ships in the Pacific—the battlefield was being fought in scientific achievements, but it was nevertheless very much a conflict and very much seen as a war. By that point, everyone knew that World War II had been won because science had given the Allies a significant advantage, and most believed specifically that the competition with the Soviets was going to be won on the battlefields of science. The first satellite (Sputnik) had been launched by the Soviets in 1957. All it did was beep, but it orbited the earth for three weeks, taunting Americans with the danger of falling behind scientifically. People in 1961 were scared.

Even though many of the subjects in Milgram's experiment were very tense and anxious during the experiment and several were extremely upset to discover that they had been willing to punish people to the point of danger, almost all of the subjects, when surveyed afterward, said they were glad they had participated. This suggests that the subjects (the teachers) were willing to put themselves through a deep unpleasantness for science.

My question, which has never been answered and probably never can be, is *Would Milgram's subjects have shocked themselves?* While no one to my knowledge has tested this question directly, in a 1981 study Daniel Batson and colleagues asked subjects if they would be willing to trade places with the learner and found that 63 percent (28 subjects out of 44 tested) were willing to trade places.[3] It is true that the Batson et al. experiment was slightly different than the original Milgram experiment. For one thing, the subjects were undergraduate psychology majors. For another, the shocks were identified as low intensity, and the reasoning given for the learner not wanting to continue was a personal history experience that made getting shocked particularly traumatic. Nevertheless, the Batson et al. study suggests that the issue is more about sacrifice for science than about following orders. I'm not sure that we should see Milgram as being about "obedience to authority." I think we need to view Milgram as being about what a person will do in service of what they see as the greater good.

For the Greater Good

Milgram's experiment is an interesting conflict between sacrifice and teams. Pushing through (continuing to shock the learner) means being willing to sacrifice one's compatriot for the greater good of science and the Cold War being fought at the time. Not pushing through means being willing to sacrifice the experiment to ensure the safety of your fellow humans.

I think this makes Milgram's experiment a very good model of the ordinary Germans committing crimes against humanity as death camp guards. They almost certainly actually believed that the people being killed were not innocents, even though they were unwilling to say so in court once they had woken up from their crimes and realized what they had done (and that anti-Semitism was not going to be a winning defense). It is important to say that the murdered millions were innocents and that what the death camp guards did was morally wrong. Understanding their reasoning does not release them from their criminality.

This is, in part, the dual nature of the *science of morality* that I am trying to get at in this book. The science of morality explains how human nature is willing to commit violence for what it perceives to be the greater good. (That's the *descriptive* [*is*] side of the question.) But the science of morality also says that we can measure the moral value of a system by how close to an assurance game it has driven us. (That's the *prescriptive* [*ought*] side of the question.) Understanding that moral target, we can address the question of whether some moral systems are better than others. (Checking up on your compatriot's safety is good. Genocide is bad.) We will come back to this question of why morality is more than *parochial altruism* (cooperation within a group and xenophobia between groups) when we address the problem of moral relativism (chapter 18). But, here, I'd like to look at the descriptive side of the question. I want to ask, *What processes changed whether subjects would continue through to the end or would reject the experiment earlier?*

Changing the Question

Milgram ran a whole series of experiments, testing different conditions. Fascinatingly, Milgram did not do a lot of gender testing. The victims were always men, and in most of the original experiments, the subjects were men as well. He did one experiment with women as subjects (punisher, providing the shock) and found similar results to the men in his experiments. Replication studies that used both men and women have found similar levels of willingness to shock.[4] Nevertheless, we can learn from the conditions that Milgram did test. Milgram reports eighteen different experiments in his book, including changes in the relationship between the teacher and the learner (punisher and victim) and the relationship between the teacher and the experimenter. The short statement is that the clearer it was that there was a problem, the more likely the subject was to stop the experiment and refuse to continue. But also, the closer the punisher was to the victim, the harder it was to punish them, and the less sure the experimenter seemed, the easier it was to say no.

In the original experiment, there was no feedback from the actor/victim. All the teacher/subject had to go on was the statement of the experimenter. In this condition, 26 people out of 40 (65 percent) went all the way to the maximum shock, and the rest typically stopped somewhere in the extremely intense shock range. Milgram then added in voice feedback, where the subject heard the victim cry out and complain. This turned out to have little to no effect. Some subjects stopped earlier when the victim first cried out (in the strong shock range), but 25 people out of 40 (62 percent) continued through to the maximum shock. However, if the subject could see the victim or if the subject had to physically put the victim's hand on a shock plate, then people were much less likely to continue through to the end. Only 16 out of 40 (40 percent) and 12 out of 40 (30 percent) were willing to continue to the end in the two experiments in which they had to physically interact with the victim.

Moreover, the more remote the experimenter was, the less likely the subject was to provide the shocks. When the experimenter left the room and provided responses only when contacted by telephone, only 9 out of 40 (20 percent) of the subjects were willing to continue through. But to me one of the most interesting effects is that when the experimenter left the room, the subjects cheated—they shocked the victim less than told to. In fact, Milgram reports that many subjects explicitly lied to the experimenter, telling the experimenter that they were providing stronger shocks than they actually were. That's a fascinating response. Here we have subjects who want to show the experimenter that they are participating in the experiment but finding a way to avoid punishing the victim as much as instructed.

Providing too-small shocks, of course, destroys the validity of the scientific experiment that the subjects thought they were doing. However, very few of the subjects were scientifically trained (they were generally citizens drawn from the local community, such as postal workers, dockworkers, deliverymen, and retail salesmen), and understanding that complexity may well have been beyond them. Thus, it is not surprising to see the subjects trying to have it both ways—shocking the learner less while still allowing the scientist to do his experiment.

Finally, most of the discussions about these experiments concentrate on the fact that a subset of people were willing to proceed to the end of the shock sequence, even as the victim screamed and even as those screams went silent. But Milgram describes these subjects as constantly questioning the experimenter, asking for reassurance that the experiment was safe, and he describes them as showing stress, tension, and anxiety at their actions. They looked for support in making their decision. In the original experiment, the experimenter (with authority) assured them it was safe and most continued. However, when peers questioned the experiment, only 4 out of 40 (10 percent) agreed to continue. Milgram conducted one experiment in which there were two experimenters, and one questioned the first experimenter's authority. In this case, none (0 out of 40) of the subjects continued to the end.

Milgram did one experiment in which another teacher came in and said, "If you won't do it, then I will." People who had decided to stop the experiment fought against that new teacher. They ripped wires out of the machine and prevented the new person from giving shocks. Once the decision had been made, people were willing to fight for their beliefs. This speaks, again, to the greater good the subjects had shifted to. Once the subject decided that this experiment was not part of the greater good, they were unwilling to let it continue. It was not just about their own moral soul (*I don't want to do it, but I won't stop you*) but rather about whether the experiment was permissible at all (*Not only am I not going to do it, but I won't let you either*). Milgram really had a handle on a moral key.

What we have in Milgram is a perfect example of moral conflict—on the one hand, the subject wants to do what's right for their country (which was, remember, at war, albeit a "Cold" one) and for science (even to this day, the vast majority of Americans have strong positive views on science[5]), yet on the other hand, the subject does not want to harm their fellow humans. We can see the conflict in the tension of the subjects and in their questions and complaints throughout the experiment. How do people resolve that conflict? By looking to their fellow humans for guidance. When the experimenter was a credible scientist and projected confidence in his decision, people tended to follow his lead, but when the experimenter was questioned, particularly by confident alternatives, people tended to reject that authority and stop the experiment.

In a sense we can view the Milgram experiment as an example of the trolley problem. The trolley problem, originally proposed as a thought experiment by philosopher Phillipa Foot, entails a situation in which you have a trolley heading for five workers but which you can divert to kill only one person, saving the five. The trolley problem asks, *Is it morally acceptable to divert the trolley?* The Milgram experiment asks, *Is it acceptable to sacrifice the learner for science?* While, certainly, shocking the learner in the Milgram experiment is not the same as sacrificing the worker to save five others, were these ordinary Americans playing their role in the great war?

It turns out that the trolley problem provides an important entry point into the neuroscience of morality. It shows that how we ask the question changes what answers we find—because how you ask the question changes which decision systems drive behavior.

THE TROLLEY PROBLEM

The trolley problem is a morality thought experiment that addresses a very ancient question: *Is it morally right to kill one person to save many?* In the now classic formulation, you are a traffic controller in a train yard. A train (trolley) with faulty brakes is hurtling toward five workers on the track. But there is a switch before the spot where the workers are working, so you could, if you act quickly, divert the train away from the workers. However, there is a worker on the other track. So diverting the train would kill that one worker. Should you divert the train? Most people (not all) say yes—killing one person is terrible, but it's better than letting the train kill five.

Would it matter if the one person were someone important or talented or skilled in some special way? Are some people worth more than others? While our current culture denies this on the surface (that is, it would be uncouth to say that some people are worth more than others), of course there are, in practice, vast disparities in how our modern societies treat individuals, and many earlier cultures would have unquestionably said that some people are worth more than others. The point of the trolley problem is not to provide an answer to this question but to clarify the question itself.

We are going to use the trolley problem to get at the descriptive side of the science of morality, particularly as a step toward mechanism. Thus, the question is not *Are some people more valuable than others?* but rather *Do humans treat some people as more valuable than others?* It is also scientifically interesting that the historical trend is very clearly moving toward a general equality—toward the answer being that all humans have an equal right to life, liberty, and the pursuit of happiness and that no person is more valuable than another. We can see this in the expansion of rights across groups and the expanding emphasis on providing individuals equal voices in government. We will come back to this observation, its causes, and its implications in the second half of the book when we look at how institutions and policies interact with our neuropsychology to change the games we play.

But let's assume, for the sake of simplicity, that the six people are all equivalently important/talented/useful/worthwhile and that by whatever equality definition you believe in, they are all worth saving. But what if you knew one of the people involved? Would it matter if you knew the one person? Would it matter if you liked or disliked the one person more than the five other workers? Knowing the one person who would be killed by diverting the train almost always makes a big difference.[6] Why does knowing the person make a difference? This leads us to the most fascinating point of the trolley problem that will open up a new perspective on the role of neuroscience on morality.

The trolley problem—version 2: You're on a bridge above some train tracks. There is a large man next to you leaning over the edge, watching a trolley hurtling toward five workers. The trolley is far enough away that if you pushed the man over the edge in front of the trolley, it would kill him but stop the trolley, thus killing one person to save five. (For the sake of argument, we'll assume that there's a reason you can't stop the trolley by sacrificing yourself. Maybe you're physically smaller than the other person on the bridge?) Do you push this guy over the edge? Almost all people say, "No, of course not." What is fascinating is that people will generally say yes to throwing the switch but no to pushing the man over the edge. What is different about pushing the man rather than throwing the switch?

In general, the closer one is to the action and the more visceral the action feels, the less likely one is to sacrifice the one person to save many. For example, if you ask the question in a person's second language, people are more likely to sacrifice one to save many, even in the scenario of pushing the man over the edge. Presumably this is because making decisions in a second language forces one to think more logically and less viscerally.[7]

We can measure what parts of the brain are active during a decision using a tool called *functional magnetic resonance imaging* (fMRI). The specifics of fMRI are complicated and beyond what we need to get into here, but we can say that fMRI detects when a part of your brain is working hard.[8] Thus, for example, if you are looking at a painting, your visual cortex shows activity, while if you are listening to music, your auditory cortex shows activity. fMRI is essentially safe and can be used on people even while they are doing tasks (playing games, selecting actions). So we can use fMRI to measure which parts of the brain are working hard when someone is making a decision about the trolley problem.

As noted in chapter 7, different decision-making systems depend on different neural circuits. What has been found in the trolley problem is remarkable—people making the logical decision (throw the switch) activate one set of neural circuits, while people making the emotional decision (don't kill) activate a different set of neural circuits.[9] The first circuits are the same as those activated when we make deliberative decisions, while the second circuits are the same as those activated during instinctual or Pavlovian decision-making.

This distinction between deliberative and Pavlovian systems is the same as that seen between choosing sweet foods (instinctual/Pavlovian) or healthy foods (deliberative) and the same distinction seen in impulsivity tasks like the classic marshmallow task. Moreover, these same distinctions differentiate sharing from selfish decisions in the ultimatum game and other tasks.[10]

The reason that people come up with two answers in these two scenarios (yes, throw the switch; no, don't push the person in front of the train) is that they are not actually equivalent. One of the intrinsic underlying starting points of our instinctual/Pavlovian system is "don't kill people you know." But, deliberatively, one could argue that it is better to sacrifice one person to save five. The push-the-person scenario is visceral enough to activate the instinctual/Pavlovian horror at killing someone while the throw-the-switch scenario is separate enough to activate the deliberative, more logical system.

THE REAL TROLLEY PROBLEM: KILLING AT CLOSE RANGE

Some researchers have expressed concern that the trolley problem is an artificial story that is not realistic. But we've already seen that humans are willing to sacrifice others for the greater good (the Milgram experiment), so maybe the trolley problem isn't so artificial. We can take the question of soldiers in wartime as a very real trolley problem: *Are you willing to kill for your country?*

As is well known from both a very large fictional literature about wars and from sociological studies of actual soldiers, much of the willingness to fight is to protect the small group of close friends that bond together in response to danger. (This is an example of coalition defense.) However, often the best way to protect your group of close friends is to get them out of the way of fire, not necessarily to kill the enemy, and studies find that it can be hard to get a soldier to actually kill an enemy. Studies in World War II estimated that only a few soldiers ever actually killed anyone. Studies of recovered rifles from the Battle of Gettysburg found that most of them were still loaded, and many were multiply loaded without firing so as to make them inoperative. There is strong historical evidence that many soldiers fired over the heads of the enemy in battle or may have missed on purpose.[11] There is also strong evidence that getting a soldier to actually kill an enemy takes specific training and very strong evidence that killing changes a person psychologically.[12]

As we saw in the Milgram study, it was much easier to shock the person who was not immediately present than it was to shock someone directly.

Similarly, it is easier to kill from a distance than it is to kill at close range.[13] Note that this does not imply that killing from a distance does not change someone. Incidents of post-traumatic stress disorder (PTSD) exist throughout the military, including in pilots as well as infantry and even including drone pilots killing from bases far away from the battlefield.[14]

One can create distance through both physical distance and through social distance, which is, of course, the point of much of the propaganda demonizing and dehumanizing the enemy in war. Getting someone to actually kill in a given situation often requires training. One can train them to do so logically by convincing them that it is necessary for the greater good, thus making their deliberative systems willing to take the action as a dark step on a path to a better end. One can train them through constant drilling of procedural systems until they react quickly to a given stimulus. Or one can train their Pavlovian systems so that they release the fight response in a given situation. Of course, whatever the process, these changes make it hard to go home again. Moreover, deciding to train people in these ways has important moral consequences that we will explore in depth throughout the rest of this book.

SUDDEN SACRIFICES: RUNNING INTO DANGER

We often talk of moral actions as a question of how much you would give of yourself to help another. The trolley problem is talked of in terms of the moral question of whether you would sacrifice another person to save five people, but the real moral question is what would you sacrifice of yourself? ("I've solved the trolley problem," says the demon Michael in the third season of *The Good Place* as he gives the last ticket to freedom to Eleanor and stays behind to hold off the bad guys.)

Interviews with people who have taken extreme risks to save others rarely find careful contemplation before the action. They often say, "It was just the right thing to do" or "I don't know why I did it, I just did it." These are signs that the other systems, instinctual/Pavlovian or procedural/habit, are in control. (This is part of why I insist that we include these noncognitive systems in our definitions of ourselves! If you are the "press secretary" explaining what another self has done, if you are a "superficial shell" controlling only a small portion of your actions, if these noncognitive actions are "zombie processes,"[15] then we remove the good that these systems do along with the bad. I would rather claim that all of these systems make up the person you are, especially since you can change what actions these noncognitive systems take in the moment by practicing and learning in the preparation.)

In his book *Humankind,* Rutger Bregman quotes a bystander diving into a freezing river to save a mother and child in a drowning car as saying, "It was like an instant reflex. Car in the water, that can't be good." (Remember that Pavlovian actions are prewired like reflexes and happen quickly without cognitive planning, but the Pavlovian system uses different neural circuits and requires more complex computations than reflexes. In the folk psychology that many people have, all of these noncognitive decision systems get lumped together, so we should not be surprised to hear someone call a Pavlovian response a "reflex.") Similarly, Abigail Marsh cites several cases, including the famous one in which then mayor of Newark Cory Booker ran into a burning building to save the people trapped inside, which he described as "You just decide to jump in."

The idea that these responses are driven by the instinctual/Pavlovian system should not be taken as suggesting they need to be learned; remember, the whole point of the instinctual/Pavlovian systems is that you don't need to learn to run from a lion. In chapter 9 we will address the question of the *intrinsic goals* that underlie these motivations, but here I will note that it is clear that responding to a person in danger is an instinctual response. In a direction opposite that of killing at a distance (where it is harder to kill someone in close proximity), the closer someone is physically, the easier it is to dive in to save them.

In fact, this is what is generally seen in bystanders helping. The more dangerous the situation, the quicker it is recognized as an emergency, making it more likely that bystanders will dive into the fray and help.[16] There is a myth that bystanders do not intervene in dangerous situations, which comes in large part from the infamous story of Kitty Genovese. Kitty Genovese was a young woman in New York who was attacked and murdered in the alley by her apartment building one night in 1964. The story written up at the time claimed that her neighbors ignored her screams for help and that some even watched from their balconies. However, subsequent in-depth reporting that actually looked at what happened found this description of the death of Kitty Genovese to be a falsely constructed myth. Several of the neighbors did call the police, the neighbors did come help, and Genovese in fact died in her friend's arms.[17] Quantitative, scientific studies of real events (using modern analysis methods applied to traffic cameras) find that bystanders generally rush toward crises to help, not away from them. People who dive in to help talk about their decisions using language that suggests an intuitive (instinctive, Pavlovian) decision-making process rather than a deliberative one.

That being said, many people do freeze in response to danger (freezing being a part of that repertoire of actions available to the instinctual system). While there are people willing to dive into an icy river to save a drowning child or to rush into a burning building, not everyone takes that action. (This is why we respect those people as heroes—they did something we hope we would do.) But there are people with jobs that require them to proceed into danger, jobs where hesitation can mean the difference between life and death. So how can we get a firefighter to enter that burning building? How can we get an emergency medical technician to continue work despite the danger of a collapsing bridge? Or a soldier to cross that firing line to bring back a fallen comrade? Moreover, we'd like them to do it safely, using the procedures most likely to get both them and the person they're trying to rescue home alive and well. We can't just trust their instincts. The key lies in that word "procedures."

THE ROLE OF HABIT: PROCEDURAL DECISIONS

Firefighters, police officers, and soldiers face sudden dangers and moral decisions every day. While early theories, like early philosophy, thought expertise would entail the development of better and better reasoning to address those decisions as individuals gained experience, actual studies of experts in these fields found that it was not deliberative decisions driving these fast responses in the face of immediate dangers but rather sudden and fast decisions. It turned out that these fast decisions were being made differently from Pavlovian decisions. Pavlovian decisions are instinctual actions that can be released in unlearned situations and that you can learn to release in new situations. Instead, these experts were recognizing patterns they had seen over many years and using these well-learned pattern-recognition processes to release arbitrary, well-practiced action sequences.[18] In the decision-making language we are using here, expertise is about training the procedural/habit system to do the right thing.

So what is the role of habit in moral decisions? The procedural system tends to be agnostic to the morality of its actions. Remember, the instinctual/Pavlovian action-selection system is going to take prewired species-important actions and learn when to release them, which means one could evolve intrinsic goals that make one more likely to succeed at the assurance game (while not getting played for a sucker). The deliberative system can reason through logic to get to a moral or immoral decision. But the procedural system can be trained to learn to recognize any arbitrary pattern to release any arbitrary action chain.

Habits take practice, but we can train ourselves to act correctly in given situations. The danger, however, is that these habits can bleed over into other situations where they are inappropriate. For example, much of military training entails learning skill-set habits to react quickly to events in a specific (and often violent) manner. Sometimes it can be difficult to react differently outside of the war itself. (Don't kill the guy at the bar who threatened you—he's just drunk. Don't become violent with your spouse when they disagree with you.) This has been known to be a problem for thousands of years.

Unlike the Pavlovian and deliberative systems, procedural decisions are inherently amoral—they are learned action chains to be taken in a given situation. But this means that procedural systems are culturally dependent. If we build a cultural training system that teaches our procedural systems to behave in a moral manner, we will find ourselves doing so when procedural systems are in control. This suggests, as I've been hinting at throughout this book, that understanding how decision-making systems interact with moral questions has important consequences for how we design our institutions and policies.

MORAL REASONING

All of the decision systems learn—they just learn in different ways. Pavlovian systems learn when to release which instinctual action. Procedural systems learn action chains to be used in response to a given situation. Deliberative systems learn to predict the consequences of one's actions. A rule-based system needs to learn what rules to follow. Even the support systems (like the situation-recognition component) learn. (The situation-recognition system learns what factors to pay attention to in the world.) Changing the way we teach ourselves and each other changes how our action-selection systems respond to the world around us. It can make us more or less likely to cooperate; it can make it easier or harder to toe that moral line.

One of my favorite stories is the "One You Feed," which is about training your decision systems. The story is that you have multiple sides within you, seen as two dogs—one wild and violent, the other loving and caring. The idea, however, is that while you might not be able to control which side reacts to a given situation in the heat of the moment, you can over time strengthen one side or the other so that when the time comes, the side that responds is the side you want. A good way to think about this parable is in terms of two possible Pavlovian (instinctual) reactions to a sudden situation, whether one uses a negative social response (fight/flight) or a positive one (cooperate/help). Remember, the Pavlovian system is learning the correct situations in which to release these species-important actions. In the moment, if the Pavlovian system is in control, you're not deliberating, you're responding. How you respond depends on the one you feed.

As a specific moral example of this, we can look at some of the recent terrible cases of police attacks on citizens, particularly the murders of African American men and women. The causes of these murders are broad and complex, and not all of them are going to follow the specific example we will go into here. Moreover, there is an important sociological question of whether these incidents have increased in number or whether we are more aware of them because of our increased ability to record them and share them on social media. On the one hand, this racial violence is certainly not new, particularly from law and government entities, and there were many such incidents during the Jim Crow era and the civil rights movement and earlier. On the other hand, there has been a dramatic increase in the militarization of the police since the 1980s (often attributed to the drug war but also related to the development of SWAT units taking up military surplus) and an increase in training that puts police at odds with the citizenry rather than as a part of the community. (Note the group dynamic moral implications here.) There is strong evidence that these changes have had a negative effect on both the police and the citizenry. This is the consequence of a science of morality. We can talk about why these policies aren't working, and we can find specific evidence-based changes to improve the situation.[19]

As a way of exploring this issue of interactions between training, decision systems, and morality, let's take the specific case of Philando Castile, who was killed by a jumpy cop not a mile from my house. Castile, his girlfriend, and her child were pulled over for driving with a broken taillight. Castile had his hands on the steering wheel when a young police officer, Jeronimo Yanez, asked him for his license and registration. Castile explained that he had a firearm (he was licensed to carry it) but that he was not reaching for it. Yanez jumped and fired seven shots at point blank range. Four of them hit Castile and killed him. We know the whole incident because we have both the police camera from the police car and because Castile's girlfriend started live streaming the event on Facebook immediately after the shooting.

No one denies the sequence of actions. Yanez was faced with a conflict between decision systems: an instinctual system that feared being in a potentially dangerous situation interacting with an unknown person who he knew had a gun and a deliberative system tracking a situation that had not yet escalated to actual danger. Yanez's reactions on the police camera, postshooting reactions on the livestream, and statements made immediately after the incident suggest that he jumped the response. The repeated "Fuck!" that can be heard is clearly from a person trying to deal with a situation he doesn't want to be in. It is the reaction of someone who has acted with a Pavlovian decision system and now realizes that he has made a mistake that will be impossible to undo.

Our understanding of decision-making systems can help reveal what happened here. Yanez had just completed a militarized training workshop in which he was repeatedly told that his life on the streets was in constant danger, that every time he stepped out of his vehicle he was entering a war zone, and that he had to be ready to kill or be killed, none of which were appropriate perspectives for this situation.[20] His Pavlovian/instinctual system was on high alert and reacted, incorrectly, to the situation. He'd fed the wrong dog.

Importantly, not all police shootings are due to Pavlovian failures. Daunte Wright was killed because of a procedural mistake when the police officer pulled out and fired her gun instead of her taser. (There is also the question of whether she needed to tase him in the first place, which is a policy implementation morality question that also interacts with our decision-making systems.) And George Floyd was killed because of a deliberate decision to deprive him of oxygen. Our understanding of decision-making systems can help us address these three situations and hopefully inform policies that can help us eliminate their occurrence.

Whenever I drive by the memorial for Philando Castile by the side of the road (which I do every day), I always wonder what might have happened that day if Jeronimo Yanez had gotten training that told him he was part of a community instead of part of an occupying force in a war zone. I always wonder if such training could have saved Castile's life. And, while I cannot speak to the specifics of Yanez's mind post incident, these kinds of events typically haunt the perpetrators and create PTSD and moral injury (chapter 11). They also lead to high levels of guilt and suicidality, particularly when they occur in the heat of a moment with strong negative (and regretted) consequences.[21]

THE SOCIOLOGICAL ENVIRONMENT

All three decision systems interact with our sociological environment to drive moral and immoral actions. We can deliberate to identify what decisions will be best for us, and if we can recognize that we are part of a cooperative community, then we can recognize that cooperating within that community will be a better choice. Procedural systems can be trained to release well-practiced action chains at the right time to make us more likely to choose the cooperative action and improve both our own outcomes and our interactions with our communities. Instinctual systems have evolved over time to create species-important responses that, on average, improve the propagation of genes into the next generation, and because individual success in social humans is better in groups of cooperators, those species-important actions include socially positive components.

What scientists have found through morality experiments (the ultimatum game and the trolley problem are experiments in morality) is that many of the behaviors we think of as "moral" (sharing, fairness, an unwillingness to kill) arise from the instinctual/Pavlovian/emotional systems rather than from the deliberative/planning/cognitive system. We see this same dichotomy in other aspects of moral behavior. For example, Tania Singer and her colleagues measured fMRI activation in third-party punishment, in an experiment where they allowed a third party to spend resources to punish people who had behaved unfairly in a variant of the ultimatum game. They found that punishing people for misbehaving created positive emotions in the instinctual/Pavlovian circuits.[22]

Note that I don't want to be Panglossian or a Pollyanna about this—the instinctual/Pavlovian system is creating situations of parochial altruism, sharing within the group and xenophobia between groups. But we can change who is part of our group, and we can create nested and interwoven groups that lead to competition between groups without violence or genocide. And we can create structures that bring out deliberative or procedural processes when we find ourselves relating to strangers whom we need to find a way to cooperate with. Importantly, all three systems learn. Deliberative systems depend on an understanding of the consequences of our actions. (*If I do this, that will happen. Is that good?*) Procedural systems depend on learning to recognize patterns and release practiced actions. (*Practice doing good works.*) Pavlovian processes learn when to release an instinctual action. (*It depends on which dog you feed.*) How we train these systems will change how we react to the world around us, including our moral and immoral actions. Importantly, however, these learning processes interact with our *intrinsic goals.*

9 INTRINSIC GOALS

All animals have species-important motivations that they seek out. In neuroscience and psychology, these are called intrinsic goals. These fundamental desires have implications for morality and suggest the first steps toward changing the structure of the world we live in.

My colleague Shmuel Lissek has a video game task he runs his subjects through in which they are a farmer and have two paths to their fields, a long path and a short path.[1] If they take the long path, crows sometimes eat their crops before they get there, and their imaginary farm family goes hungry. If they take the short path, they sometimes get a physical shock. Amazingly, people will risk that physical shock to get to their in-game, imaginary crops so that their in-game, imaginary family won't go hungry. Shmuel is studying anxiety. The short path has a symbol on it that predicts the shock with variable levels of accuracy. People with anxiety disorders generalize the shock prediction across symbols more than people without anxiety disorders. This is important research that will help us improve the lives of people with anxiety disorders, but what I keep coming back to is that the experiment works at all. I am fascinated by the fact that this experiment works even though the family is imaginary. To me, this is a remarkable observation. People are willing to accept physical pain simply because they don't want an imaginary family to go hungry. What is even more remarkable is that whenever I tell people about this, no one seems surprised but me. In psychology, the story that Shmuel tells his subjects (feeding your family) is called a *frame*, and it captures something deep and important within us—we don't want to let our family down.

People (and all animals) have evolved as a means of propagating genes on to the next generation. This was Darwin's amazing realization—that if you have two simple statements that (1) traits are passed down from parent to child and (2) traits produce differential success in survival and reproduction, then that differential success will create a process that selects for some of those traits over others. Over time, incredible changes can occur. Of course, Darwin didn't know about genes, but genes define traits and Darwin, like any observer, knew that children showed combinations of traits from their parents. In fact, Darwin built a lot of his argument in *On the Origin of Species* on observations of pigeon breeding, which was apparently a very popular pastime in Victorian England.

So evolution selects for things that are going to lead to an increased likelihood that your genes will continue into future generations. But this turns out to be a very complex problem to solve, with lots of subtleties, particularly in a social species such as humans. It is not merely a question of having enough sex to produce a child. There are questions of which genes to combine yours with. There are questions of whether it is better to support your own children as compared to your siblings' children (remember that your siblings share your genes, so helping your siblings' children also propagates your genes into the next generation). It is not enough to simply make children. Human children begin life completely vulnerable and unable to care for themselves. Raising a child to an adult capable of passing their own genes on to the next generation takes decades of work. It is not enough to simply make and raise children; your genes are evolved to maximize the number of genes in later generations all down the line, so your children must be capable humans when they grow up as well. This is a very large calculation that no one would ever be able to do explicitly.

Imagine if every action you took was based on an explicit calculation of whether it was more or less likely to increase your genetic success. Even if that explicit calculation was possible to do (which I sincerely doubt), it would take a very long time (hours, days, longer?) to make any decision. A simple question such as whether to get out of bed or to drink a cup of coffee would be impossible to answer.

Instead, we have evolved *intrinsic goals* that are correlated with genetic success. Classic examples are things like sex—most people have sex because they like it, not because they are trying to procreate. (Yes, some couples are "trying to have a baby" and are having sex in order to procreate. Nevertheless, I believe my statement that the vast majority of human sex occurs for other reasons is pretty uncontroversial.) Another classic example would be eating and drinking—most people eat food and drink water because they are hungry or thirsty, not because they will starve or dehydrate if they don't. Being attracted to certain people or being hungry or thirsty are intrinsic goals that we pursue for the goals themselves. These goals, however, are things that, in general, are correlated with survival and successful propagation of the genes into the next generation.

But . . . the fact that we chase these intrinsic goals themselves leads to some less-than-optimal behaviors. Dogs chase cars because they are similar in size and speed to the big game animals wolves evolved to chase down. We can find hundreds of other easy examples at both psychological and neurophysiological levels, such as the urge to watch images of unattainable beautiful people of the gender of one's desire or our obsession with sugary food and drink. Why do people eat too much sugar? Because when our behavior was evolving, sugar was a rare commodity that provided a lot of energy and was particularly found in the fruits we needed to eat.[2]

Part of the point of this chapter is that measuring the likelihood of genetic success is very complicated. But it is not the point of this chapter to argue that the point of our lives is to propagate our genes. Even people who don't have children contribute mightily to the human endeavor. The point of this chapter is to note that the intrinsic goals we have within us produce behaviors that lead to satisfying lives that on average succeed genetically without having to do that calculation every time. Identifying these intrinsic goals is going to be important when we start asking how to change the rules of the game (our institutions and policies) to make us more likely to cooperate in the assurance game.

Let's come back to the example of dogs chasing cars. Dogs chase cars because dogs have evolved to hunt down large game animals in packs. Cars are about the size of a typical moose and often move at about the same speed (about thirty-five miles per hour, or fifty-six kilometers per hour), and chasing a car can provide satisfaction to a dog that misses the wild. I am not suggesting that dogs are mistaking cars for moose any more than humans jogging are mistaking the mile around the suburban park for a hunting run. (Humans are some of the best long-distance runners in the animal world. Several studies have suggested that physiological changes allowing humans to run long distances enabled early hunters to run game to ground by tracking and chasing them over many days.[3]) But, in the same way that a dog can derive satisfaction from chasing a car down a suburban street or a human can derive satisfaction from the invigoration of a multimile jog, an undergraduate playing Shmuel's farmer game can derive satisfaction from taking a shock to feed their imaginary family.

Animals have evolved intrinsic goals that they pursue. Those intrinsic goals define fundamental species-important behaviors. *What are the intrinsic goals that make humans different?*

ON THE UNIVERSALITY OF INTRINSIC GOALS

Before we proceed, it is important to quickly digress to note that although I call these intrinsic goals, I make no claim as to whether they are genetically wired into our brains or whether they are consequences of the worlds we grow up in. (Nature versus nurture is a false dichotomy. Consequential phenotypes are an interaction of genetics and the environment.) Humans are evolved to grow up in social groups. Human infants are not viable for several years without some parental care. And a human who grew from young childhood without any other humans around would be so far from normal as to be deeply problematic.

In this chapter, we are addressing the descriptive side of human behavior, asking the question *What do humans do in these situations?* As such, we cannot ignore the negative side of this descriptive story. I made the case in the first part of the book that there is a prescriptive side to morality as well and that we should not take observations about what humans do as a moral imperative. To do so leads to what is known as the *naturalistic fallacy*, the idea that whatever we observe humans actually doing is morally right. This leads to moral relativism. I am not going to make that assumption; in fact, in later chapters I will explicitly argue against it.

Similarly, I do not want to get into questions of where these goals come from. I am not going to speculate on the evolutionary psychology of these goals, which often devolve into just-so stories that are hard to justify and require tremendous (often unfounded) assumptions about the social structures of ancient hominids. Instead, I note that these intrinsic goals are good general descriptions of modern humans across many cultures, including Western, educated, industrialized, rich, and democratic societies;[4] other industrialized societies across the world; extant subsistence farming societies; and extant hunter-gatherer societies. Moreover, these intrinsic goals are well attested to throughout the arts, including in modern literature (books, movies, television) as well as in earlier literatures across civilizations (from *The Epic of Gilgamesh* to Shakespeare, from *The Iliad* and *The Odyssey* to the *Ramayana* and the *Mahabarata*, from *The Tale of Genji* to the Bible) and throughout mythology, which often provides remarkably human descriptions of the motivations of the gods.

Finally, these are generalizations. Your mileage may vary. While I would argue that these are pretty good generalizations across human societies, they do vary across societies and across individuals. For example, most humans (not all humans) enjoy the taste of sugar, but Americans and Europeans have very different preferences for the sweetness of desserts (many continental Europeans find American desserts far too sweet for their taste). Even humans who enjoy the taste of sugar are able to reject that taste. Diet varies greatly across individuals and cultures. Individuals can decide to eat low-sugar diets. In fact, the current treatment for diabetes (a dysfunction in the ability to secrete the insulin necessary to process sugars and carbohydrates) is to make food decisions using the logical, deliberative system through explicit calculations of how much insulin to provide oneself artificially.

These intrinsic goals should not be seen as inevitable descriptions of any individual. Rather, they should be seen as foundations that we can use when we start to try to build tools to change the games we're playing.

The point of this chapter is to recognize that these are fundamental aspects of humanity. In the same way that dogs like to chase big things that run about thirty-five miles per hour, humans have a set of intrinsic social goals they are drawn to, which happen to be well evolved to help us solve the assurance game within our local community group but which we need to expand on to move beyond that community.

We can identify a few particularly interesting intrinsic goals that humans have related to social community and morality. (This will, of course, be an incomplete list of the intrinsic goals of human beings.) First, humans have an intrinsic goal of fairness, both of fairness that involves themselves and also, importantly, others. Humans do not like seeing other humans being unfair. (To quote the hero-to-be Steve Rogers [*Captain America*] when asked if he wants to join the army in World War II to "go kill Nazis": "I just don't like bullies. I don't care where they're from.") Second, there is a lot of evidence that humans are social creatures and that loneliness is devastating to one's health and well-being. Third, humans have a need to be a part of a larger story, to have a sense of purpose. We find that purpose by defining a group and finding a role to play within that group. Moreover, we express fundamental within-group controls that prevent selfishists from infiltrating our groups and fundamental between-group behaviors that encourage the success of our group against others.

Importantly, not all of these intrinsic goals are necessarily positive. For example, humans seem to have intrinsic goals of wanting to mete out punishment and revenge for slights and insults. Humans also have intrinsic goals of wanting to be seen as powerful and successful, which can manifest as a desire for hierarchy (as long as one is on top of it). And humans have intrinsic goals that separate their own group from others, creating xenophobia. We will need to understand both the positive and the negative intrinsic goals in order to begin to build cultural tools to change the game we are playing.

Let's take these one at a time.

An Intrinsic Desire for Fairness

Going back to the ultimatum game, in which one player chooses how much to share with another and the other can either accept or reject the deal (rejecting the deal means no one gets anything), humans react with emotion and anger to a small offer and often reject that offer.[5] Interestingly, humans are generally willing to take any offer if told that the other "player" is a random number drawn from a computer but unwilling to take small offers when told that the other player is a person. This suggests an important aspect to the frame: when we interact with other people, we demand that they play fairly.[6]

Many animals, particularly primates, have a keen sense of fairness. If one monkey gets a grape (a tasty and sweet treat) while another gets a slice of cucumber (bitter and less preferred), the monkey that gets the cucumber will throw it at the experimenter and scream angrily. Importantly, that same monkey would have accepted the cucumber if there was no other monkey getting grapes.[7] Similarly, human children will tally up exact accounts of who got more cake or a longer time on the computer or more chores to be done with the cry that parents of siblings know everywhere—"It's not faaaiiiir!"

As with all of the discussions in this book, I do not want to be naive about this—the fact that one child decries the lack of fairness of the distribution of rewards does not mean that the distribution was unfair. However, most parents react differently to valid cries of unfairness and invalid cries, which is exactly what you would expect in the arms race of group dynamics. Children test their abilities to game the system, and parents have to teach their children that gaming the system is not possible.

Notice the importance of emotions in these discussions. As noted in chapter 7, emotional responses are a key indicator of the Pavlovian/instinctual system, and in fact, studies of the ultimatum game find that, like the trolley problem, different reactions to the ultimatum game access different neural systems.[8] Unfair deals activate neural systems associated with emotional Pavlovian systems, such as the orbitofrontal cortex and anterior insula. Rule-based decisions activate more deliberative systems, such as the dorsolateral prefrontal cortex. Encouraging the deliberative system makes one more willing to accept an unfair deal that benefits one in a utilitarian way, while encouraging the Pavlovian system makes one more willing to sacrifice one's winnings to punish an unfair deal.[9]

Nothing hurts worse emotionally than thinking you were safely part of a group and finding yourself betrayed. Whether it be a small group (such as a betrayal by a cheating spouse), a medium-sized group (ostracized by people you thought were your friends), or a large group (discovering that your leaders are actually Quislings selling your country out to the enemy), these betrayals produce nauseating emotions of anger and despair. The concept of betrayal depends on the notion that we are part of a community and that it is a fundamentally anti-human response to break that community bond.

Punishment and Revenge

We saw earlier that under simplified assumptions, second-party punishment can provide a stable system that supports cooperation in a repeated prisoner's dilemma (which makes it an assurance game). This was the tit-for-tat discussion we saw in the Axelrod simulations (chapter 4). Not surprisingly, as social creatures, humans have intrinsic desires to punish insults, slights, thefts, and betrayals. As we saw, the problem is that these simplified assumptions make relying on pure dyadic punishment dangerous and insufficient.

What does one do in response to a mistake? Is there a way back from betrayal? We saw in our discussion of social control (chapter 6) that modifications of tit for tat allowing forgiveness and redemption can provide for a return to cooperation after a defection. However, even in these situations humans often express a desire to punish the offender, in what is called *retributive justice*.[10] Moreover, it is very difficult to keep that retribution balanced. Dyadic punishment easily escalates, whereby the response to retribution is increased retribution, spiraling into a feud between the two parties. Literature is filled with such stories. As we've seen, one solution is to develop third-party punishment processes that can bypass this retribution cycle.

One complexity is that third-party punishment is usually decided upon and meted out by individuals within a community (such as a judge or jury in a court) who often see retributive punishment as a goal in itself. Which person those individuals meting out the punishment empathize with will often determine which side they take in a conflict, such as a trial. This explains both over- and underpunishment. (We will address the question of empathy in detail in chapter 10.) In his book *The Punisher's Brain*, Morris Hoffman describes how he, as a practicing judge, must force himself out of a quick-response mindset when deciding how much of a sentence to impose on offenders in his courtroom. Over time, he says he has learned a set of tricks that work. (Take time before responding. Write notes carefully. Identify specific questions to answer for each defendant.) These tricks are all tools that shift decisions from Pavlovian to deliberative decision-making systems.

In general, these tricks are good ways to remove bias by focusing attention on the specific issues that underlie the goal of a given decision. For example, when hiring a new person, identify the characteristics you want that employee to have before meeting any candidates, then explicitly address each of those characteristics for each person. In his book *Thinking Fast and Slow*, Daniel Kahneman has an excellent description of how these processes work in actual practice. In *The Punisher's Brain* Hoffman does note that with experience he has found himself able to reach reasonable, correct decisions even when thinking quickly, but that is the development of procedural and pattern-recognition systems.[11] For our purposes, it is clear from Hoffman's descriptions that the problem is Pavlovian responses that need to be shifted to either deliberative or procedural systems for fair punishments.

Another solution is to shift from *retributive justice* to *restorative justice*, which creates a goal of repairing the harm done, rather than punishing the offender. As described eloquently by Hoffman, justice has three goals: specific deterrence, reducing the likelihood of an individual committing the offense again, general deterrence, reducing the likelihood of another individual committing a similar offence, and restoration, undoing the harm created by the crime. When we are constructing sociological structures, we will want to address all three of these goals.

The complexity is that people have an intrinsic desire for punishment as a means of changing behavior. I will always remember the commencement speaker at my college graduation, who told the class of engineering students that we should follow the science wherever it leads and implement the right solutions based solely on the science and then, not five minutes later in his speech, ranted that he didn't care if the science said that drug use was a disease and that drug treatment centers worked better than jail—those monsters should be locked up because they saw the anti-drug ads, they knew better, and they should all be jailed forever. (There is extensive evidence that jailing drug users is counterproductive and that drug treatment is a much more effective process of returning drug users to society.[12])

Being Part of a Group—the Opposite of Lonely

Humans have the need to be part of a group. Enforced solitude is devastating to the human psyche and creates dramatic personality changes that are often difficult to recover from.[13] Loneliness has been identified as a key health factor with a causal role in a variety of health conditions (such as heart disease, stroke, dementia, and other disabling factors). In general, lonely people have a higher mortality rate than people with a support structure.[14]

Note that it is important to differentiate loneliness from solitude. Loneliness can be defined as having fewer social relationships, of more superficial quality, than one likes. First and foremost, this means that one can be lonely even when surrounded by other human beings. Second, it means that how lonely one is will depend on the relationships one desires, which, of course, will change as a function of one's mood and other circumstances. Certainly, a quiet walk in the park can help one reset, calm down, and regenerate from the difficulties in life. A writing retreat is not loneliness. I can tell you from personal experience that sitting on the edge of a mountain lake after a long silent hike and just listening to the birds can be wonderful. Nevertheless, most humans (at least sometimes) enjoy the company of others and want (and need) a support structure of other humans to function normally.

We will see that resilience to psychological trauma is associated with a strong support structure and that post-traumatic stress disorder (PTSD) is associated with a lack of a support structure to help process trauma. While traditional human social structures are based on a tribal group of kith and kin, as well as key family structures, humans also seek out their own groups for support. One of the most interesting concepts here is that of a "found family"—a set of people who will support you in your times of trouble. A true friend is the one who "when you have to go to them, they have to take you in."

Xenophobia

While within-group interaction can provide support and resilience, there is a negative side to our intrinsic goals of being in a group—that of xenophobia. Dozens of books have made the case that humans are evolved to show *parochial altruism*—cooperation within a group, but xenophobia outside of it.

We cannot deny that xenophobia exists. Nor is it hard to explain why it exists. The key factor underlying the observation that altruistic groups do better than selfish groups (which was the whole key to the first part of the book) is competition between groups, which leads to the prediction that such processes encourage cooperation within groups and competition beyond them. We may have evolved to be a particularly friendly species, but we are preferentially friendly within our groups, not between them.

This is the key to coalition defense, which is the idea that when faced with enemies who have banded together to attack us, we need to band together to survive. Peter Turchin, in a series of well-argued books (particularly *Historical Dynamics*), points out that empires form at the boundaries of cultures as individual tribes, bands, and clans join together to repel invaders. Coalition defense comes from the simple observation that numbers do matter—when attacked by a large group with asabiya, one needs to respond with a large group with asabiya. Internal conflict is dangerous when facing an enemy with asabiya. The danger of xenophobia comes from the fact that these boundaries are often defined ethnically. It is easier to bond with people you are familiar with. This means that providing experiences with diversity to increase familiarity can reduce these ethnic boundaries and make it easier to cross those boundary lines.[15]

The success of groups playing our assurance game (tax game, public goods game) requires that groups with asabiya do better than groups that don't. It requires competition between the groups such that success or failure occurs at a group level, like a sports team winning or losing. It does not actually require violence or xenophobia. That is going to open up the very important possibility of competition without conflict.

A Role in a Greater Purpose

Humans have a need not just to be part of a group but also to have a role within that group. They want to know that they are playing a part that is helping. Whether we are talking about sports teams, family life, a military unit, or even society itself, humans want to play a role within that group. The strength of human society is that of specialization. I would much rather pay someone who is an expert roofer to build a better roof on my house than to try to figure it out myself, while that roofer would much rather I work on discovering new things about the world that will help us improve our lives.

Of course, these roles interact with the fairness questions. I have been writing much of this book during the COVID-19 pandemic. During this pandemic, a number of important societal services have had to continue (grocery workers, trash collectors, mass transit drivers), and workers performing those services have been declared "essential." Interestingly, many of those workers have been paid very low wages over the years because they have never been historically recognized as essential. There is growing support for increasing the wages of those workers now that we are recognizing their importance.

So how do we motivate someone to play that essential role? The key, of course, comes from having a purpose—the idea that we are working together toward something greater than ourselves, a purpose that we cannot achieve alone but that our group, our team, our humanity can. We want to be a part of something greater than the sum of its parts. Moreover, we want the respect that comes from playing our role in that greater group.

For example, let's take the cases of striking sanitation workers in 1968, such as those in New York City and Memphis. Throughout, the complaints were about respect—the civic government did not respect the sanitation workers because they didn't think they were important to society. But after nine days of garbage piling up, it became clear that the sanitation workers were absolutely critical to the community. What was important was the idea that the sanitation workers played an integral role in the New York City community. Similarly, the signs worn by the sanitation strikers in Memphis did not say, "I demand more pay." They said, "I am a man."

Status, Success, and Hierarchy

High-status humans are more genetically successful than low-status humans. This is so obvious as to be almost an accepted truism. However, while general observation, qualitative anthropological descriptions, and quantitative sociological studies have all found this to be true on average, we have to be careful about what we mean by status.

And, of course, we need to separate statements of description from statements of prescription. The fact that studies find that, historically, high-status individuals usually have more children who live to have their own children does not mean this is the way it should be. Moreover, most of these studies have a remarkably male-dominated viewpoint. While it is true that there are differences in reproductive success as a function of status among women as well as men, the differences are often much smaller.[16]

Humans are primates and many studies have made analogies to primate dominance hierarchies when talking about human status. But even there, one needs to be careful. The actual social structure of primate communities is not as simple as a linear dominance hierarchy. For example, male and female hierarchies interact in chimpanzee, baboon, and vervet monkey societies. Bonobo societies are generally much more driven by dyadic interactions of mutual cooperation and trade. And these hierarchies are driven by much more than physical violence or strength. Social interactions are often as important, if not more so.

Studies of baboons and chimpanzees have found that the superficial dominance hierarchies of who-can-beat-up-whom does not actually translate into reproductive success or, for that matter, whom the troop listens to. The most physically dominant male baboons are the young bucks trying to make their way up the hierarchy, but it is the older males whom the females seek out, and it is the older males and older females who decide when the troop moves to a new foraging ground. Hierarchy in baboon troops may be better described through female family relationships than through the males, likely because males are the ones who transition between tribes, while female relationships can be stable for decades, or generations. The complex stories of the nonhuman primates are beyond the scope of this book, but there are dozens of wonderful studies looking at these relationships.[17]

And, beyond that, human status is far more complex than other primates'. For one thing, humans don't have a single hierarchy. Humans exist in a vast network of overlapping groups. This means that who is high status depends on the context. The high-status boss in a workplace may or may not be the one to make decisions at the community social or the town baseball game or in the volunteer fire department. Who is in charge can change in an instant when an expert is needed (think of the doctor who arrives on the scene of a medical emergency). In our pirates example from chapter 5, the ship's surgeon outranked the captain in medical questions, and it was not always the captain who led the boarding party in a battle. Beyond these complexities, human societies rarely have a single dominant alpha who is in charge. Governments are run by coalitions of groups, and the asabiya of that group often defines the status of that group relative to others.

Importantly, not everyone wants to be in charge. In fact, status is not necessarily about being captain of the ship. Being in charge is often a precarious position, giving one a target on one's back for others to make their name by taking you down. In general, the precariousness of status is stressful, as is the responsibility of the consequences of one's actions. Of course, being low status itself is also extremely stressful, in large part due to the micro- and macro-aggressions that low-status individuals are subject to. In short, one has to be very careful about assuming generalities as to what defines status, how status interacts with society, and what the effect of status is on one's role in society. Perhaps we should say that what people want is a little R.E.S.P.E.C.T.[18]

While one can study nonindustrialized hunter-gather/foraging tribal societies and find complexity in the male and female hierarchies, it is a leap to make assumptions about inherent individual goals from these other societies. They are just as much complex societies as are modern industrialized societies, which themselves can vary tremendously across cultures and groups. We do not actually know in what social environments humans evolved nor how interactions between society and individuals have changed humans over time.

Nevertheless, we can identify some basic universals, particularly about status and respect. High-status humans are more likely to be given opportunities, including financial, educational, and sexual opportunities, are more likely to get additional resources, are more likely to be listened to and respected in conversations, and are more likely to have influence over others. All humans have an intrinsic goal of being high status and of being respected. Status in humans is highly context dependent—that is, your status depends on your group. Someone can be high status in one group and low status in another. Status often depends on expertise and experience.

There are many ways to achieve high status in societies. One can gain status through talent (think a rock star musician or a Nobel Prize winner), through family (think a rich scion), through hard work (grit and determination), or through sacrifice and bravery (diving into a river to save a child,

Medal of Honor winners), but one can also gain status by putting down others (by dehumanizing groups that are different, by limiting opportunities of others, or by genocide). Sociological studies often talk about social mobility versus social inertia—how similar is the status of children when they grow up to that of their parents? Caste structures limit opportunities, while mobile social structures provide them.

Importantly, this means that there are positive and negative ways to gain status. One can gain status by increasing the cooperation opportunities within an assurance game or by forcing others into the losing corner. Social structures will change both what status means and how one gains that status. That means the intrinsic goal for status can be a tool that we can use to guide behavior, providing awards and status recognition for the behaviors we want to encourage.

Other Intrinsic Goals

This list of social goals is, of course, incomplete. There are social goals of family success, as well as that of friends and students. (Many academics see their students as descendants. In my own field, one of the popular sociological pastimes is to examine *NeuroTree*, a wiki-like family tree of who trained who. This can be useful because students pick up the ideas and perspectives of their mentors, and knowing someone's "family tree" can help one understand where a colleague is coming from.) Many researchers have suggested that human social groups evolved by extending intrinsic goals of family success to friends and colleagues beyond the family.

We have not yet talked about nonsocial intrinsic goals, such as hunger, thirst, or the more complex desire for variety in foods and flavors. Most people don't like eating the same food every day and want variety in their diet. There are human-general desires for certain natural scenes (a field by a river, mountains in the distance). The specifics of what aspects are truly universal across all humans and which are culturally derived is beyond where we need to be in this book. But, in order to build our new science, we will need to study how institutions and policies interact with these intrinsic goals as they change ourselves, our societies, and our behaviors.

In a sense these institutions that we need to create are cultural tools. We can think of cultural tools as being like any other tool in the human arsenal. Humans are the only species who have successfully settled every ecosystem on Earth. Not only are civilizations found in jungles, temperate climes, savannas, deserts, and arctic ice; we now even have bases under the ocean and in space. Humans have thrived in arctic climes where one needs to hunt and fish in order to survive, in farming communities across temperate breadbaskets, and in jungles where vegetation is common but food is scarce. While our bases under the ocean and in space are not currently self-sufficient, is there anyone who doubts we could make them such if we put our minds to it?

Other species create tools that have been genetically encoded (such as the shell of the caddisfly larva, which is built out of stones and mud) and tools they have learned to create (such as jackdaws and crows building hooks or otters cracking sea urchins on a stone). Other primates create tools and learn skills that they copy and modify from each other (such as the chimpanzees of Gombe fishing for ants with sticks, the chimpanzees of the Tai Forest cracking nuts with stones, or the monkeys of Koshima Island washing sweet potatoes in salt water). But these are limited cases. Humans have thousands of tools and skill sets passed on by culture. One of the major strengths that has allowed humans to conquer every ecosystem is that humans have cultural transmission beyond that of any other species.[19]

Cultural transmission allows those skills to improve as they are transmitted. With cultural transmission, different communities have different tools and skills available to them, whether these are how to use different tools to get food from hidden sources or how to prevent selfish behavior within one's community. The advantage of cultural transmission is that it builds on itself. An individual does not have to reinvent the wheel. While it can take thousands or millions of years to evolve a new tool (such as the caddisfly shell) and change would be very slow if each individual had to figure out how to create a tool on their own, cultural transmission allows us to build on each other's discoveries. Every so often, I find it fascinating to stop and look around at all of the cultural tools we have built, from the computer I am typing this on, the internet that is storing my text as I type it in the cloud, the house providing me a roof over my head, the clothes I am wearing, the knowledge of how to roast and filter coffee to wake me up in the morning, to the language that I am writing and you are reading.

Humans take this cultural transmission to a level beyond that of any other animal.[20] In part this is because humans are imitators extraordinaire. In a remarkable experiment, human children were asked to open puzzle boxes (boxes that required complex knobs, locks, pull rods, and other operations to open). Some of these operations were not actually necessary to open the puzzle box, but if the children were shown a demonstration that included those operations, they did them anyway. However, when children were shown the puzzle box and asked to determine the opening process sequence on their own, they did not do the extra operations. They were more efficient when they did not have these inefficient people to learn from. Humans learning from another person follow the extra steps. Other primates don't.[21]

Even in small groups, humans imitate each other. Many studies have found that two people talking with each other tend to begin to imitate each other's physical behaviors, smiling when the other smiles, tilting their heads together.[22]

Moreover, humans are the only animals that actively teach. While a mother cat may bring partially killed prey for her kittens to learn to hunt and chimpanzees will hand a built ant-fishing stick to a child, there is no evidence that any other animal actively corrects mistakes made by their students in the way that humans do.

Furthermore, humans like rituals, and they bond through ritual, such as marching in formation or singing in church. Groups performing a ritual together become more willing to participate in group activities and more willing to work together as a group. But these effects can be negative as well. One of the most cited and replicated psychological studies is that people will often misreport reality if everyone around them says something untrue. Experiments have found, for example, that people faced with two lines drawn on a page, one longer than the other, will report the shorter line as the longer one if they are in a room where everyone else reports that incorrect statement first. An important question here is how much the individual is simply reporting a false outcome to go along and how much perception has actually changed. What is interesting is that, like the Milgram experiments discussed in chapter 8, not everyone followed the group in these experiments (remember, 35 percent of Milgram's subjects refused to increase the shock to the danger zone, even in the original study). It was easier to disagree with the group if there was another member disagreeing, and social group construction played an important role in the willingness to deny the perception. Again, in a remarkable similarity to the Milgram studies, individuals showed anxiety, hesitation, and uncertainty when faced with this conflict between their individual perception and that of the group.[23]

This group perception problem can be seen in larger-scale issues as well, particularly when an understanding of the situation depends not only on immediate perception but on the interpretation of complex inputs (such as deciding what to do in a business plan or when prosecuting a war). This is sometimes known as the *groupthink* problem. Groupthink obviously reduces the effectiveness of a group—it diminishes the ability of the group to learn from all individuals. We can come back to the parable of Salman Rushdie's *Haroun and the Sea of Stories*. The Guppees argue among themselves until they reach a consensus, while the Chupwalas shut down all dissent. The Guppees win because their group is stronger. They have more asabiya (chapter 5). We can create policies that encourage discussion, such as the classic "brainstorming rules" that start with a first pass where ideas are thrown out without dissent or judgment and only discussed later, or policies that go around the table, asking everyone to contribute, or the classic "Yes, and . . ." rule of improvisational theater.

Culture, laws, religion, and government are all ways of changing the rules of the game. Some of these rule changes affect the payout matrix, as we saw in the first chapters of the book. If you take too much from the common resource, the community will punish you, making it less valuable for you to cheat the common resource. Of course, how the community punishes you changes what opportunities there are for redemption. As we've seen (and will see), these journeys to redemption depend on third-party community norms, second-party interactions, and first-party (self-driven) aspirations, as well as how these constructions interact with intrinsic goals.

These rule changes interact with our decision-making systems in important ways. For example, over the last thirty years or so, federal and local police have been providing "pop-up" shoot-or-don't training in which a realistic cardboard cutout of a person pops out in a realistic setting, and the subject has to make a split-second decision to fire or not.[24] The theory, of course, is that if you are in a combat situation, reacting quickly can be the difference between life and death. Except that actual studies of combat find that soldiers facing that kind of situation generally hesitate, often allowing one of them to flee rather than die.[25] Note how one party fleeing provides the opportunity for redemption in a way that killing and dying does not. More importantly, police are not supposed to be in combat situations; they are supposed to be helping provide third-party community control (chapter 6). The problem, as we now know from our discussion of decision-making systems (chapter 7), is that this pop-up training experience is going to activate and train the officer's procedural system, which, to remind you, recognizes patterns and releases an action chain. That means that the pattern recognized is going to be based on superficial factors, such as skin color and clothing, which is, of course, deeply problematic and has cost many an innocent person (even children) their lives.

When I think of this kind of training, I always think of the darkly funny scene in the movie *Men in Black* where the key recruit being tested (Officer Edwards, soon to be Agent J, played by Will Smith) makes shoot/no-shoot decisions that differ from those of the other recruits. Rewatching that scene, it is even darker and more poignant than I remembered. All of the alien pop-ups are snarly, scary-looking aliens just going about their lives. Agent J is the only one who sees it—because he identifies with each one. "He's just working out. How would I feel if someone busts me up while I'm working on the treadmill?" "He's not snarling, he's sneezing." And yet he himself makes a snap judgment about little Tiffany, assuming that the advanced quantum physics textbooks she's carrying are such a mismatch with her age and size as to imply that she is sufficiently dangerous to justify killing her on the spot.

10 EMPATHY AND TRUST

Empathy is a key factor that humans use to engender trust with each other and a key factor in solving the assurance game.

Empathy allows one to understand another person and walk a distance, at least mentally, in another's shoes. In colloquial English usage, empathy can be contrasted with sympathy. Sympathy entails understanding another's feelings but not necessarily feeling them oneself. A host of other emotions are involved in social interactions, but empathy plays a particularly important role in the social interactions we are talking about. In that scene from *Men in Black* mentioned at the end of chapter 9, Officer Edwards (Agent J) empathizes with the aliens ("How would I feel if someone busts me up while I'm working on the treadmill?") but not little Tiffany, which leads to his choice to shoot her and not the others.

The assurance game explains why empathy is critical to the question of morality. In the assurance game, the best option is to do what the other person is going to do—cooperate if they are going to cooperate with you, but defect if they are going to defect. Empathizing with the other person primes you to take the same action, to cooperate when they do, and to defect when they do.

This is what makes the assurance game different from the other games we looked at in chapter 2, like the prisoner's dilemma and zero-sum games. In the prisoner's dilemma, no matter what your partner does it's always better for you as an individual to defect, so you don't actually need to guess what your partner is going to do. In a zero-sum game, you only want to know what your partner is going to do in order to maximize your reward against them. (Think of the D-Day invasion, which we noted was a form of the matching pennies game. The Germans needed to know where the Allies were going to land, at Normandy or Calais, so they would know where to put their defenses. All that mattered was determining if the army poised to attack Calais was real or illusory. No empathy needed.) But the assurance game is different. Empathy can help solve the assurance game.

Technically, one could solve the assurance game deliberatively by figuring out what the other player is going to do and deciding the best thing to do in that situation. But, in practice, humans generally drive cooperation through empathy more than through deliberative decision processes.[1]

Neurophysiologically, we know that empathy depends on the ability to imagine what it would be like to be that other person. In mammalian brains, imagination uses the same neural systems as perception. What I mean by this is that the part of your brain used when you perceive something (visual cortex for images, auditory cortex for sounds) is also used when you imagine that thing (visual cortex for remembering a painting, auditory cortex for remembering a song). If you imagine a motor action, like a golf swing, your motor cortex becomes active.[2]

Empathy is the imagination of another's emotions.[3] If you imagine how happy your toddler will be chasing bubbles on a spring day, the part of your brain that represents that childish happiness becomes active. If you imagine how someone will feel when you betray their trust, the same parts of your brain that would become active when you are betrayed become active. Studies have found that these include not only the neural components involved in Pavlovian decisions (such as the amygdala) and emotional representational areas, such as the orbitofrontal cortex and the insula (two areas near the bottom front of the brain) but also the cingulate cortex and somatosensory cortex, which are active when you feel physical pain. Being rejected socially literally hurts.[4] Betraying someone you empathize with can create the same physical pain as getting betrayed yourself, which makes it difficult to betray that other person.[5]

If we go back to the *Golden Balls* stories that we looked at in chapter 2, Ashleigh and Ben got to the cooperate-cooperate corner by empathizing with each other. In their video they each talk about how they felt betrayed in previous encounters with other people and how great it would be to successfully split the pot this time. In contrast, Ella and Michelle never actually tried to understand each other and ended up playing different games. In their video, Ella and Michelle seem to be talking past each other. (To remind the reader, Michelle said that she was going to share because it was the right thing to do, while Ella said that she was happy with her choice.) Ella chose to steal once she knew Michelle would split, as if she were playing a zero-sum game. Once she knew what Michelle would do, she knew she could safely choose steal and win and be happy with her choice. Michelle, on the other hand, decided to choose split no matter what Ella chose, which is not the right way to win an assurance game. (To be fair to Michelle's perspective, she was giving Ella a chance to do the right thing and considered Ella's failure to split with her a moral loss for Ella.)

It is important, however, to recognize that just because someone empathizes with another doesn't mean they won't betray them. From her facial expressions, I suspect that Sarah empathized with Steven but took his money anyway. I'm not sure that Michelle could have actually done anything to drive Ella to choose split, unless she took an out-of-the-box action like Nick did, which demonstrates that one can also win at an assurance game independent of empathy. Nick drove Ibrahim to play split by changing the rules of the game and sticking to his guns, not by empathizing with Ibrahim nor by asking Ibrahim to empathize with him (Nick).

TELL ME YOUR STORY

So what creates empathy? How do we empathize with another person? If empathy depends on imagining what it is like to be the other person, then it likely depends on one's ability to understand the other's story.

Humans are narrative creatures. Stories have been a key factor in our social lives for millennia. Telling stories and imagined narratives are one of the few true universals across all human societies. While it is unclear whether other animals also have stories to tell, it is unquestionably true that humans define themselves narratively. This is why we can redefine ourselves by retelling our own story, changing our view of how we see ourselves by reframing our actions in light of a different view on our history.

Empathy depends on understanding and the ability to recognize what it is like to be that other person and to imagine that other role. Thus, empathy depends on stories. Extensive evidence demonstrates that experience with diversity increases one's understanding and tolerance of that diversity.[6] Ecumenical discussions between groups can increase understanding.

This is one of the common arguments in favor of trade and interactions between groups. If we go back to our canonical trade example of our mountain and valley farmers (chapter 4), I'm sure that when our mountain village farmers take their crops to the village in the valley, they will have a chance to interact with them, to talk to them, and to hear their stories and likewise when the valley farmers carry their crops to the village on the mountain. In fact, interactions like this tend to include celebrations, food sharing, and other positive social community-building opportunities.

Of course, as our communities get larger and larger it becomes more and more difficult to hear everyone's individual story, but communities can be built on myths and stories transmitted by cultural means. One of the fascinating aspects of popular literature is that people are perfectly willing to tolerate magic, science fiction, and changes to the physical world, but they are very unwilling to tolerate changes in human behavior. We are happy to imagine ourselves in a world where the wand chooses the wizard, and kids are informed of their acceptance to a secret boarding school from magic letters delivered by owls, but what makes the Harry Potter series work is Harry's interactions with his friends and the very real consequences of the choices made by the characters. The transformation of Severus Snape from villain to tragic hero lies in our understanding of his story: his mistakes that drove away and killed the woman he loved and his desperate attempts to atone for those mistakes by saving the son she had with the man he hated.

THE NEUROPHYSIOLOGY OF TRUST

Neurophysiologically, subjects playing trust games show empathy with the other player through reflection of the other player's emotions. (*What would they feel if I acted this way?*) As we noted earlier, one of the most important neuroscientific discoveries of the last thirty years is that imagination accesses the same parts of one's brain as perception. Similarly, when one imagines how another person is feeling, the parts of the brain that process our own emotions become active. The key brain structures involved in these trust questions are the amygdala, the ventromedial and orbitofrontal cortices, the anterior cingulate cortices, and the anterior insula. The amygdala is usually associated with Pavlovian actions, such as freezing in fear, hesitating because of anxiety, and seeking reward, but it is also keenly activated in social situations. Observing emotional responses and expressing empathy for them produces amygdala activations. The ventromedial and orbitofrontal cortices are involved in evaluating emotional outcomes. The anterior insula is involved in third-party social control, such as when a subject punishes another in a trust game, and in concerns about betrayal.[7]

A problem with one's Pavlovian systems could lead to difficulty in appreciating the intrinsic human need that connects one to other people, and a problem with emotional representations could lead to difficulties in empathy. People with these characteristics are more likely to defect than to cooperate, and when they play repeated assurance games, they are more likely to drive themselves (with their partners) into the defect-defect corner, even when playing against normal individuals.[8] Abigail Marsh points out that people with these difficulties are at the extreme tail of one side of the distribution of humans and set out to study people at the extreme tail of the other side of that distribution. She studied people who had donated organs to strangers as evidence for an individual's extreme altruism and found that those *super-altruists* showed stronger activation of emotional structures in empathy.[9]

Moreover, when she tested the range of people's social networks, she found that people with these disorders had social networks that fell off quickly, and they thought of people they knew well as strangers, while super-altruists had social networks that fell off particularly slowly, and they treated strangers like close friends. Howard Rachlin has long argued that in the same way we look at time through a discounting function (the now is important; the future far away, abstract, and less focused) and in the same way we look at space through a discounting function (the local is concrete, the far away abstract and less focused), we create social networks that decay with distance. The super-altruists were treating everyone as part of their close kith and kin, while psychopaths were treating no one that way.

We are more likely to trust people closer to us in that social structure and less likely to trust people further away. The more one empathizes with a subject, the easier it is to sacrifice for them. Abigail Marsh has found that people willing to donate organs to strangers are much more likely to empathize with individuals they barely know. Daniel Batson has done extensive studies on the relationship between empathy and altruism and has found that one's willingness to help others is closely predicted by how well one empathizes with them. Several researchers have argued that this empathic concern can arise evolutionarily through steps that begin with concern for one's offspring, then evolve into concern for one's kin, and, finally, expand into larger and larger social groups. Not surprisingly, the neurophysiology of this empathic concern is closely related to the neurophysiology of parenting and protecting children, which brings us to oxytocin.

Oxytocin

Given all of the discussion of oxytocin, both in the popular media and the actual scientific literature, we can't leave a discussion of trust and the neuroscience of choice in the social domain without talking about it. Oxytocin acts as both a neurotransmitter and a hormone. A neurotransmitter is a chemical that is released from one set of neurons and detected by another set, allowing neurons to communicate with each other. A hormone is a chemical that is released into the bloodstream to communicate with other bodily organs.[10] Oxytocin is involved in affiliative relationships between humans—when we hug each other, when we smile at each other, when a baby nurses. Oxytocin plays important roles in sex, birth, and lactation. It drives one person to partner up with another. It is even released in pet owners (snuggling with a purring cat or scratching a dog behind the ears).[11] But our story starts with the discovery that the difference between monogamous voles that mate for life and polyamorous voles that do not maintain a lifelong partnership lies in the receptors to oxytocin.[12] (Humans are more complicated than this, as we will see below.)

Let's start with the voles because that's where the current literature really took off. Voles are a mouselike species (genus *Microtus*) that live in the

American West. There are two highly related voles, the prairie vole (*Microtus ochrogaster*, living in the midwestern prairies) and the montane (mountain) vole (*Microtus montanus*, living in alpine meadows in the Rockies). Prairie voles tend to form long-term pair bonds wherein a single male-female pair mate multiple times, producing multiple litters together, and montane voles tend to form short-term partnerships wherein pairs mate and then separate. Both voles produce oxytocin during mating, but prairie voles have more receptors for oxytocin in brain structures involved in decision-making and reward, while montane voles have lower receptor densities. Changing the receptor density or efficacy of oxytocin in montane voles through genetic or pharmacological manipulation makes them act more like prairie voles and vice versa.

Current theories suggest that neuromodulators with broad impact such as oxytocin have a general signaling effect—they carry some important information signal that changes the processing in lots of different brain structures, often in different ways. Just what that signal is, however, remains unclear. Oxytocin is thought to manipulate one's bonding to another, which would be an important part of group formation and thus would have an impact on our moral experiments (those economic games we've been exploring).

Oxytocin and Interactive Economic Games

Because oxytocin is relatively safe, it can be delivered in experiments to individuals. (It can be made into an aerosol that can be delivered to the brain through the nose.[13]) A number of experiments have examined the effects of oxytocin on ultimatum, dictator, and trust games.

In normal, healthy adults, oxytocin increases sociality, increasing the amount given in ultimatum games by Player A, particularly when played against humans rather than against a computer. (To remind the reader, in the ultimatum game Player A has $20 to distribute between themself and another person, Player B. Player B can accept the offer, at which point the money is distributed as suggested, or can reject the offer, at which point the money is taken away, and no one gets anything. This allows Player B to punish Player A for offering a bad deal. The ultimatum game is contrasted with the dictator game, in which Player B doesn't get a chance to reject the offer, and each player gets the distribution that Player A decided on.) Interestingly, oxytocin did not change the minimum amount accepted in the ultimatum game by Player B. Nor did it change the amount given in the dictator game.[14]

What differs between the ultimatum donor, the ultimatum recipient, and the dictator is the importance of the other player's ability to influence the game—the ultimatum donor has to take into account the expectations of the ultimatum recipient, but the others don't. If Player A in the ultimatum game thinks Player B is going to expect more money, then it becomes important not to shirk their share. So maybe oxytocin affects Player A's offer in the ultimatum game by changing A's empathy for Player B. One common theory is that offers in the ultimatum game are driven in part by empathy and a theory of mind: How angry does Player A think Player B will be with the split? Many theories of oxytocin are based on increased attention to social cues.

In a similar vein, oxytocin increases the amount put into the trust in the trustee game but not the amount returned from the trust by the trustee. (Again, a quick reminder—in the trustee game, Player A gives money to Player B that triples along the way. Player B is allowed to return some of the money back to Player A.) Note that both of these experiments contain controls (by telling people that the other player is random or a computer, not a person) to ensure that it is not about being more willing to take risks. Oxytocin doesn't have an effect if the other player is a computer. Increasing oxytocin increases what a person thinks the other player expects to get.[15]

Although no one currently knows what the underlying mechanism is for why this happens, current theories suggest that oxytocin increases the recognition of sociality—that is, it increases the Pavlovian processes that underlie within-group sharing. As we've seen throughout this book, humans create groups through sharing, expectations of third-party punishment, and empathy.

Neurophysiologically, the human brain structures involved in both empathy and third-party punishment contain enhanced oxytocin receptors.[16] Moreover, rather than thinking in terms of a concrete in-group/out-group dynamic, we've been finding ourselves talking about social distance in this chapter—an expanding diagram of family, relatives, friends, acquaintances, and colleagues, extending to larger and larger groups. Oxytocin seems to be expanding that distribution, making acquaintances seem more like close friends or family and making it easier to empathize with them. Oxytocin helps groups form.

But things are not so simple. While oxytocin seems to pull members of our in-group closer together, it pushes members of our out-group further away.[17] Oxytocin increases sharing within the group but decreases sharing with people outside the group. Explicit tests of in-group over out-group favoritism and implicit-bias effects (where one associates positive words with people in one's group but negative words with people outside of one's group) are enhanced by oxytocin. While oxytocin does seem to increase trust and sharing and general prosociality, it does so by enhancing the in-group/out-group dichotomy, not by simply bonding all people together.

EMPATHY AND MORAL DILEMMAS

Milgram found that it was harder to shock someone you have met—someone you see as another human being, someone you empathize with. The quotes from the transcripts in his book suggest that a large part of the difficulty people found in shocking others was through their empathy with the learner. "How would I feel if I were in that position? Not great, I tell you." The more that a subject interacted with the learner they were supposed to shock, the harder it was to shock them.

We saw in chapter 8 that the trolley problem activated deliberative systems if one was far away from the action but Pavlovian systems if one was closer. Focusing on the immediate interactive effect on the individual you have to kill (to save the five others) makes one empathize with them. That makes it harder to push the man off the bridge than to throw the switch.

Not surprisingly, studies have consistently found, over and over, that one is both less willing to harm and more willing to help someone one empathizes with. This is, of course, the key to the television ads that try to encourage charity by emphasizing individual suffering (increased empathy makes one more willing to share) on the positive side, but also dehumanization on the negative side (decreased empathy makes one more willing to kill or harm).

Dehumanization is a process whereby one is made to feel that another person (or more often another group) is less than human and thus not someone to empathize with. If someone is less than human, it becomes easier to refuse to share with them, easier to torture or punish them, easier to kill them. Dehumanization is a key tool used to get individuals to fight in wars, enforce unfair hierarchies, and even commit genocide. As we will see in the next chapter, these tools create *moral injury* when people do (eventually) recognize the other as a person and reevaluate their actions in light of that epiphany.

11 TRAUMA, MORAL INJURY, AND RESILIENCE

Post-traumatic stress disorder (PTSD) gives us an entry point into what can happen when trust and empathy break down, which can occur through both trauma and moral injury. There is, however, a positive side to the neuropsychology of trust—a supportive social structure can create resilience.

Post-traumatic stress disorder (PTSD) has almost certainly been around for millennia. Some of the first official diagnoses of PTSD-like symptoms came from the American Civil War, when doctors noticed after the war that some soldiers had incidents of heart palpitations, particularly in reaction to sudden stimuli. They called this *soldier's heart* and tested for it by having the patient sit on a couch and lighting black powder underneath him (!). In the animal experimental literature, this is termed *fear-potentiated startle*, whereby a subject (usually a rat or mouse) that has been subjected to cue-predicted shocks jumps particularly high in response to a sudden presentation of that cue.[1] Basically, the mouse, the rat, and the soldier with soldier's heart are on hyperalert and respond physically to the surprising cue—with heart palpitations, sweats, and terror. Because these were most easily observable as a pounding heart, soldier's heart was thought to be a cardiology problem, but we now know that they were having panic attacks, and we now identify this as PTSD.

People with PTSD are injured, in large part because they have learned (through experience) sociological interactions that make it difficult to integrate back into society. Understanding what has happened to people with PTSD can give us insights into those sociological interactions and how changing the rules can make things better or worse for people. Also, not everyone who experiences trauma ends up with PTSD, which might give us a clue to resilience. Perhaps if we understand what factors make someone resilient, we can teach people to be resilient, or we can construct social structures to make it easier to be resilient so that people who do experience trauma can find their way back through the valley of darkness, back into the light of day, and back into their communities.

There are many kinds of PTSD and many symptoms that show up in subjects with PTSD, including *hypervigilance* (such as the soldiers with soldier's heart), the feeling that the world is dull and flat (*anhedonia*—the inability to feel pleasure), *intrusive memories* (particularly of the trauma itself), and sometimes violent reactions to casual stimuli.[2] These symptoms have been around for millennia and go by many names (soldier's heart in the American Civil War, shell shock in World War I, battle fatigue in World War II, and PTSD in the more modern formulations).

We can even see some of these symptoms in ancient literature. Assyrian descriptions of their military campaigns described soldiers unable to stop reliving memories of violent events and soldiers who saw their dead comrades and enemies.[3] (It certainly reads like they were experiencing flashbacks.) Several of the soldiers in Homer's *Iliad* are haunted by their dead, leading them to take unnecessary risks and even commit suicide. (Suicide is one of the most common consequences of PTSD seen in modern veterans.[4]) Herodotus describes a soldier who had to be pulled from battle when he became blind after seeing his neighboring soldier killed by a Persian in the Battle of Marathon (where a small force of Athenians and their allies held off an invading force more than ten times their number).[5] In the American Civil War, the best predictor of soldier's heart was the number of one's comrades who had died, not one's own injuries.[6] Other Greek authors described fear and madness that continued long after the battle, most famously, in Homer's *Odyssey* in Odysseus's inability to integrate back into society after the Trojan War and the loss of his comrades on the long journey home.

Odysseus was the king of a small island in ancient Greece called Ithaca. He had signed a defensive pact with the other kings of Greece, and when they went off to fight the Trojan War with the goal of taking Troy, he was forced to join the expedition. Ostensibly, the Trojan War was due to the queen of Sparta (Helen) running off with ("being captured by") a Trojan prince (Paris) and her husband (Menelaus) hauling after her, but Troy sat at the Dardanelles, a narrow strait through which ships must pass to get from the Mediterranean to the Black Sea. Troy held this key to the Eurasian trade route and heavily taxed traveling ships. How much the Trojan War was dyadic punishment for a sexual dalliance, a raid on a rich city-state, or an attempt to open up a blocked trade route remains lost to history. However, I would note that excuses for war can often hide economic and strategic goals.

The Trojan War lasted ten years, and afterward, it took Odysseus ten more years to reach home. On his way he lost his entire crew, spent seven years "recovering" on a remote island with the nymph Kalypso, and returned home in secret. There he finds the Ithacan nobility feasting in his house trying to woo his wife, Penelope, while she holds them off with excuses about finishing her father-in-law's funeral shroud first. I think it is important to recognize that while the suitors are portrayed in *The Odyssey* as drunk frat boys and boors, it's more likely that the nobility were trying to become king by wooing the widowed queen.

From a political perspective, one might expect Odysseus's son Telemachus to take over the kingdom, as he is now twenty years old, but psychologically, he's not ready. In *The Odyssey*, Telemachus is unwilling to declare his father dead because he holds out hope that his father, who he never knew, is still alive. This very believable literary structure is, of course, what makes *The Odyssey* still so readable even in the modern era. We can see the cinematic structure of the story: Telemachus denying that his father is dead, the nobility demanding Penelope choose a new king, and Penelope holding out hope for her husband's return, in part because she doesn't think any of these other men are up to her husband's leadership capabilities.

Remember that Odysseus had left twenty years earlier and that all of the other kings had returned home ten years earlier. To presume him killed in action was not an unreasonable assumption. When Odysseus does come home (secretly), he reacts violently to the situation and kills the suitors, even as they run away. Many people have speculated that Odysseus had PTSD and reacted more violently than needed. Moreover, while *The Odyssey* itself ends with Odysseus reunited with Penelope, further stories from the time suggest that he was unable to stay and left again to wander the Mediterranean, like a hobo, unable to settle down.

Many of the hobos at the end of the nineteenth century famously "hopping trains" were actually Civil War veterans unable to return to normal society, likely with forms of PTSD that may or may not have manifested with soldier's heart.[7] Some were probably also driven to the hobo life because of the regular economic crises (depressions/recessions) that happened every twenty years in the late nineteenth and early twentieth centuries.[8] Remember also that there were few publicly available, positive economic safety nets at the time. (This is the same time period as Victorian England, which also lacked societal support, as described, for example, by Charles Dickens.) We will see later how economic safety nets support successful communities.

It is important to note that many kinds of PTSD exist and that most people with PTSD are not violent. In fact, we can rethink PTSD in terms of the new views on psychiatry (called *computational psychiatry*) that consider psychiatric problems in terms of "vulnerabilities" in decision-making systems[9]—here in an inability of these systems to cope with trauma. Just as there are many potential ways for a bridge to fail or that can make an airplane fall out of the sky, there are many ways that our neural information processing can go wrong, making it difficult for us to handle the social world we live in. That means there are multiple conditions where damaged decision-making systems do not pull up the correct action to take in a given situation. Some of these inappropriate actions can be to remain on hyperalert (which is exhausting), to freeze or run in terror in response to harmless stimuli, or to react violently in inappropriate situations. However, it is important to note that the largest danger in PTSD is that of suicide, not violence toward others. Survivor's guilt can be an important factor driving someone to make a suicidal decision to end the story.[10]

Many kinds of PTSD can be understood in these computational psychiatry information-processing terms. I want to address three here: incorrect habits and Pavlovian responses applied to the wrong situations, moral injury and moral hazard, and unprocessed trauma due to the loss of one's social network.

Surviving combat requires a complex skill set that would often be inappropriate in normal situations. This means that people in combat need to learn new habits and make new Pavlovian associations—they need to change their decision systems to react differently.

While many people with PTSD are not violent (or are violent only to themselves), reports abound of soldiers responding inappropriately to minor stimuli with increases in domestic violence or overreactions such as Odysseus's reactions to the suitors.[11] In fact, many combat veterans talk about the difficulty in keeping their reactions in check when someone makes them angry. In chapter 7, we talked about how decisions arise from three key systems—deliberative, procedural (habit), and Pavlovian (instinctual) action-selection processes. All three systems can produce incorrect violent responses.

Pavlovian decision systems release basic instinctual and emotional responses to learned situations and stimuli. These responses include species-important behaviors such as fighting, fleeing, or freezing (defending oneself, running away, or freezing in fear). If someone has learned to respond to a sudden event (such as a gunshot) with fear, that can translate to similar situations (such as when a doctor sets off black powder under the couch you are sitting on). Not surprisingly, this can lead to the belief that nothing is safe and thus to hypervigilant behaviors, such as checking every doorway and alerting to every noise, making it hard to interact in normal social environments. In his pair of books *Achilles in Vietnam* and *Odysseus in America*, Jonathan Shay describes former soldiers who sleep with guns under their pillows or who cannot go to a restaurant because they believe that too many potential enemies are hidden within the population. Hair-trigger angry responses to small situations can be due to overlearning Pavlovian responses. Responding with anger and violence to simple disagreements can lead to domestic violence, as seen in many PTSD-related cases.

Similarly, procedural systems entail learning complex action chains to respond in specific situations (hitting a baseball, throwing a pass, and definitely the martial arts). When fighting, seconds, even parts of a second, can make the difference between life and death, so soldiers are trained to respond quickly to specific stimuli. While training has almost always entailed asking soldiers to be sure of their targets before responding, a miscategorized stimulus can release a well-practiced violent response.

But even deliberative systems can be disrupted in war. Many of the veterans in Jonathan Shay's clinic describe plans and deliberations that go astray due to changed expectations in how the world will react, like the Vietnam War veteran unwilling to eat at a restaurant in the suburban US because the restaurants in Saigon during his tour of duty were simply too dangerous. The suicides that veterans with PTSD fall victim to are often well-planned, deliberate decisions, not sudden impulsive actions.[12]

MORAL INJURY

War is filled with difficult decisions, many of which would be insane in a normal civilized world. In the Milgram experiment, people who continued to shock the other person showed concern, anxiety, and deep discomfort. We can see these emotional interactions as a conflict between decision systems and a first step to understanding moral injury. Similarly, in the utilitarian decision in the trolley problem, subjects divert a trolley to kill one person to save five (see chapter 8). A deliberative decision process, weighing options explicitly, often chooses this *utilitarian* decision (i.e., that "the needs of the many outweigh the needs of the few"). However, in that utilitarian decision you have killed that one person in your community. You have taken an action that goes against the assurance game and against your intrinsic goals: an action that feels fundamentally immoral. Simply put, you have sinned. Moral injury is, in a sense, third-party punishment applied to oneself.

Michael Graziano suggests in his book *Consciousness and the Social Brain* that we evolved mechanisms to infer agency in order to understand the probabilistic nature of personality in others but then applied those predictive cognitive processes to ourselves. Graziano argues that self-application of those prediction abilities is consciousness. I am not going to discuss the question of consciousness in this book, but it is worth noting here that this same idea of applying third-party regulation to ourselves may explain moral injury. Graziano's observation is one of the reasons that first-person aspirational goals work. Those aspirational goals are factors we'd like to see in others, and through these self-reflection mechanisms that have evolved in humans, we apply them to ourselves. Moral injury arises when we find a deep mismatch between the moral character we believe we should have and our actions. Treating moral injury requires finding paths to redemption and forgiveness, applied, not to others, but to ourselves.

One of the best takes on moral injury is in the movie *Courage under Fire*, which is a heartrending tale of postconflict attempts to accommodate difficult decisions made under duress, whether it be Lieutenant Colonel Serling's drinking and unwillingness to see himself as a hero for his actions in the tank battle that he won, Ilario's or Altameyer's desperate retreat into drugs, or Monfriez's suicide. As noted above, these responses are commonly seen in response to PTSD.

Interestingly, the moral injury side of PTSD arises both from individuals faced with specific incidents due to Pavlovian actions (killing an enemy soldier at close range) and from recognizing that utilitarian decisions have real consequences. For example, one of the most evocative images from the Vietnam War was that of nine-year-old Phan Thi Kim Phuc running naked down a road with napalm burns. That image changed the viewpoint of many people back in the US and helped end the Vietnam War. Americans stopped seeing the war as just a conflict between governments and realized the real consequences of bombing a distant population. (Notice the similarity to the trolley problem and the importance of empathy here.)

One of the most interesting consequences of writing a book like this one is that the fact-checking required leads me to follow up on mythological stories. I had always assumed that the girl in the photo had died of her injuries, but in fact both the soldiers in the photo and the journalist taking the picture provided immediate first aid to her and the others in the village. They took her to an aid station and then to a hospital, and she survived (although she needed years of treatment). She now lives in Canada.[13] Even in war, people step up and help each other.

Returning to civilized life after these forms of moral injury requires extensive processing and the acceptance that one can be redeemed. Although most of this chapter has talked of PTSD as a consequence of war-related trauma, these same symptoms (flashbacks, anhedonia, hyperalert behaviors, anxiety) are seen in people with traumatic experiences outside of war zones (including things like car crashes, earthquakes, terrorist attacks, and physical and sexual assault, as well as from long-term, pervasive trauma such as child abuse or domestic violence).[14] Interestingly, many of the traumas that lead to PTSD are psychosocial ones. These are situations in which trust is broken, and one no longer feels safe within one's social group. Importantly, victims often feel isolated, embarrassed to admit what happened, and unsure of where to turn for supportive help.

Processing these injuries requires a social support structure. In a remarkable study, Judith Pizarro and her colleagues found that the best predictor of soldier's heart in American Civil War veterans was not whether the soldier was injured nor the amount of combat or violence experienced but rather the number of other soldiers in the unit that were killed.[15] This is a well-known phenomenon termed *survivor's guilt*, and it is a key factor in many aspects of PTSD. An excellent literary example of survivor's guilt can be found in *Flashpoint*'s episode "Behind the Blue Line," in which a suicidal soldier leaves behind a picture of his unit (all of whom had been killed but him) with the text "SYOTOS" on it—"See you on the other side." This devastates the fellow veteran who tries and fails to save him.

Many of the PTSD-related behaviors (hypervigilance, potentially violent behavior, quickly driven Pavlovian responses, apathy, and anhedonia) all diminish one's ability to work within a social structure. They make it hard to trust anyone. A soldier with soldier's heart is alone. And as we've seen, that loneliness can be deadly. However, the fact that the loss of a social network is so devastating suggests a potential positive alternative—that a robust social network can provide support and resilience.

People who experience trauma can find themselves having difficulty interacting with society, whether it be sudden acute trauma (such as a physical or sexual assault), moral injury (taking some action one deeply regrets or, for that matter, not acting when one wishes one had), or long-term trauma (such as being abused as a child). But there are journeys out of that trauma, and moreover, some people are better able to accommodate those difficulties than others. I'd like to end this chapter on a positive note by asking what makes some people resilient to trauma and what we can do to encourage that resilience (aside from the obvious choice of trying to do our best to reduce the occurrence of trauma in the first place). Resilience is the ability to weather a storm, to continue in the face of darkness.

There is a large psychological literature on resilience asking how an individual's past enables one to handle negative experiences, such as a bad performance review (*Why do some people respond by improving themselves, while others fall into depression?*) or the loss of a loved one (*Why do some people find an easier path out of debilitating grief than others?*). Experimentally, we can ask why some nonhuman animals weather the stress of predictable and unpredictable physical shocks or maternal separation better than others. This literature also asks how to create recovery in those who have experienced trauma.

What this literature finds is that while genetics and neural vulnerabilities are key factors in making some individuals resilient, other important elements include learning from small but manageable amounts of stress and the availability of social support. Is the loss of social support the reason why the loss of one's comrades was the best predictor of soldier's heart in the American Civil War? These companions not only would have been a socially tight-knit group formed in the crucible of war but would also have been a key support structure absent when the survivors returned home because most of the units were formed from towns and communities.

An additional key to resilience is the ability to place the crisis within a bigger picture of meaning and importance. A friend's father once told me that he never talked about World War II (where he had served in Europe) because he had lost his soul there, but he could live with it because he had sacrificed his soul for a necessary cause. We can see, once again, how the questions of asabiya (chapter 5) and the insight of seeing the Milgram experiment as about sacrifice for a greater good (chapter 8) seem to be keys to this puzzle. These are the same questions of community construction that we talked about in the first part of the book.

Primates in general are social creatures, and recovery from trauma in primates often depends on social interactions. This is key to John Bowlby's seminal work on attachment, particularly between parents and children. Harry Harlow's devastating maternal separation experiments found that the monkeys could recover if an older animal comforted and held the heartbroken infant.[16] And in chimpanzees, grief and recovery can be found in the company of friends.[17] Likewise, human infants in neonatal intensive care units are more likely to survive when they have daily physical contact.[18]

One of the most interesting discussions I have been having with my colleagues in psychiatry is that of *compensation*, whereby one system is used to accomplish tasks normally handled by another. But in order to use compensation correctly, one needs to know the strengths and weaknesses of each system. So, for example, while it might be difficult to deliberate in the midst of a crisis, it might be possible to set up times to plan contingencies and prepare for those contingencies deliberatively during the downtimes. (Know where the space suits are before the hull breach. This is why flight attendants go over the safety procedures before the plane takes off.)

This deliberative construction of a quick, procedurally available alternative is the purpose of suicide hotlines, Alcoholics Anonymous sponsorships, and putting the phone number for Poison Control on your fridge. The idea is that during sober and nonsuicidal times, you plan a very simple process that can be followed even during a panic. In the crisis, you will not need to think about the consequences of calling the hotline or your sponsor or the Poison Control center—only that it is the right thing to do in the moment. Then that call can start a process to cut off the dangerous path (the suicide attempt, the drinking, the panic when your kid has ingested something dangerous) and provide a new path forward.

Another option is to take stock of the intrinsic goals we examined in chapter 9 and use social support and meaning to increase resilience. Resilience depends on a sense of agency and one's expectations of one's ability to manage stress, as well as on one's social network and the meaning assigned to the disaster. But, of course, that is really about team construction and creating communities willing to work together.

This leads us to questions of institutions and policies. Morality is about changing those institutions and policies to change the rules of the game. We are not prisoners of the prisoner's dilemma. By changing the rules of the game, we can make our world a better place. We can make it easier to get through the crisis, if we can only convince ourselves that we are all in this together.

TRANSITION

12 FROM SCIENCE TO ENGINEERING

Knowledge has design consequences. Understanding how the world works allows us to build new devices, airplanes, spaceships, more stable bridges, earthquake-resistant buildings, new vaccines for sudden pandemics . . . and new societal structures.

WHAT WE KNOW SO FAR

We started our journey by looking at economic games, starting with one-shot two-player games (*matching pennies*, the *prisoner's dilemma*, the *assurance game*). We found a fundamental difference between the prisoner's dilemma and the assurance game: in the prisoner's dilemma, there is a conflict between what is best for the pair of players and what is best for each individual player, while in the assurance game, cooperating is best for both the group and for the players as individuals. Importantly, we saw that the repeated prisoner's dilemma is actually an assurance game, suggesting that being a part of a community changes the game. We were able to translate these insights into multiplayer community games (the *tax game*, the *public goods* game) and found that we could create variants that increased the likelihood that people would cooperate (such as providing the opportunity for third-party punishment). We then identified mechanisms of social control, including *asabiya*, the willingness to think of oneself as part of a group, as well as *first-party*, *second-party*, and *third-party* social control. We saw how structures that enhance self-control help ensure that one is part of a group of cooperators. We also identified the importance of opportunities for redemption and the escalation of punishments. We identified reputation as an important factor as well as the importance of social norms for its control.

We then explored how decision-making systems work and found that humans are not unitary decision-makers but depend on separate deliberative, procedural, and instinctual/Pavlovian processes. Who we are is an interaction of these decision processes, each of which interacts with the world through different computations on different representations and thus responds in different ways at different times. This meant that by asking questions in different ways, we could influence which decision system was controlling our behavior and how that decision system processed the information that controlled our behavior.

We then looked at the neurophysiology underlying moral decisions, including experiments like the Milgram shock experiment and the trolley problem. We found that different descriptions of the trolley problem accessed different decision systems in contradictory ways—subjects distanced from the consequences of the decision were more likely to access deliberative systems, while subjects close enough to the decision to make the violence visceral were more likely to access instinctual/Pavlovian systems. Importantly, we found that these descriptions had real-world consequences in the willingness to kill and in the consequences of killing. And we saw that changing policy can have life-and-death consequences, such as in the example of police pop-up shooter training.

So, to summarize, we can make it easier or harder to reach the cooperate-cooperate corner of the assurance game by changing the rules of the game through changes in social norms and social structures. Those rules interact with our neuropsychology (our decision-making systems, their computational processes, and the intrinsic goals that underlie behavior) in important ways.

THE POLITICAL ANIMAL

Aristotle called humans "the political animal" from the concept of the polis, or city. (The Greek word for city is "polis," from which we get the endings of many city names, like Minneapolis and Indianapolis, as well as words like "politics," "political," "policy," and "police.") This polis includes not only the political decisions that affect our lives but also the social milieu that we live in. A better term might be to call us "the social animal." Much of our reality is socially constructed, defined by the institutions that we have agreed upon. I am an American. In order to travel to another country, I need a document stating my nationality and that I have permission to go there. Depending on the relationships between my nationality and that other country, the documentation I need differs. I live in the state of Minnesota. I pay taxes to the state of Minnesota but not to Wisconsin or North Dakota. I am a professor at the University of Minnesota. When I walk into a classroom, there is an accepted set of roles that I and the students are expected to take. Even the concept of a class and the way that we teach college students are social constructs.

These socially constructed agreements are the rules of the game. By changing these rules, we change the game.

Not all of these socially constructed agreements are moral questions. Some of them are merely structures that we all need to agree on in order for our society to work smoothly. For example, it doesn't matter which side of the road one drives on as long as everyone in a contiguous country drives on the same side. It doesn't matter what color or shape our money is as long as everyone agrees to treat it as money. We don't want to define every single social construction as a moral code, but many of these structures have moral consequences. (For example, how should we consider social norms that make certain foods unacceptable or restrict certain types of clothing? Many of these rules are often folded into what are commonly considered moral structures.[1]) From a scientific perspective, we can ask how each of these rules changes our social interactions.

It is important to remember that the games we looked at in the first chapters of the book were simplified models to allow us to understand how changing the rules can change the game and how some of the basic neuropsychological mechanisms interact with rules and games. But, in reality, our social interactions are very complex. For example, we started by talking about the invasion of Normandy as a matching pennies game. Obviously, the invasion of Normandy included a lot more social interaction than just the decision of whether to invade/defend Normandy or Calais. Although these social interactions are orders of magnitude more complicated, the language we have built up allows us to understand these processes and the consequences of making these changes to our society.

A classic example is marriage. Many people see marriage as a social construct between two parties, and as such, it would be seen as an internal, dyadic contract. Others see marriage as a social contract between three parties—the two people getting married and a separate god—making it a triadic contract and enabling a form of reputational or third-party punishment from that overseeing party. Both of these views make sense within the context of our assurance game story as a means of making the declaration of marriage a stronger commitment than merely "being together," increasing the cost of separating. (There is good and bad in this increased cost.)

Taking our logic further, I think it makes sense to see marriage as a social construct between the two people who are getting married and the society that agrees to treat them differently. This was one of the major issues in the marriage equality movement: that marriage contains a host of both rights and responsibilities recognized not only by the couple in question but by the society that they are a part of. This idea that marriage is a social construct was key to the successful frame used to convince the US to allow same-sex couples to marry. The argument was that preventing same-sex couples from marrying made them different in the eyes of both the law and the community, which prevented these individuals from taking on the specific adult roles they wanted to take in the community. This turned the marriage argument on its head and said that legally recognizing same-sex marriage was a pro-marriage rather than an anti-marriage argument.

NORMATIVE GOALS AND THE NATURALISTIC FALLACY

Many people assume that science's goals are all descriptive; that the only goal of science is to find a better understanding of how the world works. But that's not a great description of either the work that scientists do or the reasons that scientists do the work they do. The better understanding that physics provides (through increased generalizations and symmetries) leads to better control of the physical world. Psychiatry has a goal of improving mental health, and medicine in general is about improving physical health. These are very much real goals of these sciences, and they change what aspects of the understanding matter.

Some literatures refer to the chase for understanding (descriptive science) as "pure science" and the goals of improving the world (prescriptive science) as "applied science," while other literatures refer to the first part (description) as "science" and the second part (prescription) as "engineering."

While descriptive science is only about understanding and prediction and thus does not depend on the goals themselves, what one tries to understand, what factors one looks at to improve that understanding, and even the constructs that one uses within that understanding can change with time and experience. While the answers themselves and whether a discovery is progress within science are independent of the scientist themself, what answers one looks for, what questions one asks, and what constructs one is willing to consider do (often) depend on the actual lived experience of the scientist themself. This is one of the reasons that diversity in science is important—it helps open up the questions we are going to ask. (There are other reasons to favor diversity in science as well, including arguments against leaving talent behind and arguments of fairness.)

On the other hand, changing the goals does change the outcomes of applied science and engineering. Engineering a bridge is about building a better bridge that will be more stable, stronger, and less likely to fall down. (There may be other optimization questions as well, such as cost or environmental impacts.) Our identification of morality as changing the rules to shift us into an assurance game provides us a guidepost to address these questions of means and ends when we talk about the applied science (engineering) side of the question.

Seeing moral structures as tools to solve the assurance game avoids the naturalistic fallacy of assuming that whatever groups and people do is right. Slavery and genocide are bad because they remove many people from the assurance game. Increasing equity of opportunity and outcome brings more people into the assurance game, increasing the non-zero-sum success of the cooperation corner of the assurance game. Some moral structures really are better than others at creating assurance games out of social interactions. However, while seeing moral structures as tools to solve the assurance game provides us guidance forward, we cannot assume that these answers are simple.

Solving the Assurance Game Is Not Utilitarianism

Many secular discussions of the neuroscience of morality have focused on one of two fundamental hypotheses: either the idea that morality is based on *harm avoidance* and is all about not harming another person or that it is *utility increasing* and is all about increasing the total goodness (happiness) in the world.[2] These two ideas are linked; one can extend harm-avoidance through secondary logic to try to explain support and punishment as trying to decrease harm to others. While, certainly, avoiding harm is an important part of the story, I would argue that it doesn't capture what we are seeing in the assurance game because the goal of the assurance game is not just to do no harm to your compatriots; working together actually creates increased advantages for all parties. Think of the example of the stag hunt, where cooperating gets both you and your companion a lot of meat, while defecting means you each only get a little. Similar cases can be made for the infrastructure example, such as the two towns that now have a bridge between them that the people can use to visit each other. In the more general tax game, it's not just about harm but about better systems overall that provide non-zero-sum positive rewards. Furthermore, harm-avoidance becomes complicated when we start to look at questions such as whether to shock a compatriot (Milgram) or whether to shift the tracks in the trolley problem.

As these arguments are extended, they end up in *utilitarianism*—the idea that the goal is to make the world better for everybody, measured as the sum total of "better." One of the important questions that arises in utilitarianism is who counts as "everybody" within that utilitarian goal. If everyone is equal within that goal, then one should starve one's own children to feed people in drought-stricken areas because the return on investment (the number of people one could feed for a given cost) is so much higher. No one is going to do that.

The assurance game is not a utilitarian goal. In the assurance game, you should cooperate if your compatriot is going to cooperate, but if you cooperate with a selfish player who defects against you, then you are just a sucker. In the assurance game, it is best to cooperate with cooperators and to defect against defectors. Extending the assurance game to group dynamics (the *tax game*), we want to play in a group of cooperators, but we don't want to be the only person to put money in the pot if no one else is paying their tax (if everyone else is going to play selfishly). On the other hand, I do not want to suggest that the assurance game implies that we should only cooperate with people who have the means to reciprocate that help. We will see in our discussion of institutions and policies that there are good reasons (both mesoeconomically and neuropsychologically) to protect the weak, help the innocent, and do what we can for the least among us.

One can think of the assurance game as a form of hierarchical utilitarianism—one wants to support one's own group first. Importantly, however, humans live in a complex web of groups. I have a family, an extended clan of relatives, friends, students, colleagues, a city, a state, a nation, and a world that I support. Because the assurance game is not zero-sum, the more people we can get to cooperate within our group the better we will do. Thus, in a sense our end goal is a form of utilitarianism in that we would like to extend our cooperating group to ever-larger populations. But with these nested groups, we do not fall down the rabbit hole of *you should starve your own children to feed a dozen kids in another country* because the assurance game works within communities that we have to define.

We do not want to assume that these groups are simple or straightforward. They are more a web of groups than a hierarchy. Just as we need to be careful about individuals taking excess resources within our groups, we need to be careful about groups taking excess resources within the larger space of metagroups. We also do not want to assume that these groups are predefined, stable, or based on trivial foundations. In fact, we will see that there are clear advantages to having networks of groups that tie us together in webs rather than linear chains. Different social institutions can make groups stronger or weaker, more open or more closed, and more or less likely to be playing in that cooperate-cooperate corner. Moreover, different social institutions can be designed using the knowledge we've developed (of the mathematics of assurance games and of how decision-making systems work in humans) to increase cooperation across those larger and larger groups.

Solving the Assurance Game Is Not Collectivism

Similarly, the assurance game is not collectivism, which is the idea that the goal of morality is to create a group. In many formulations of morality, moral suasion is described in terms of doing the right thing because it's the right thing for the group despite the fact that it's bad for you as an individual. While the view of morality espoused within this book includes sacrifice (a soldier dying for their country), risk (diving into the freezing river to save the drowning child), and first-person social control (don't cheat, even if no one's watching), the argument is that groups that include cooperating individuals are more likely to be playing the assurance game successfully. In the assurance game, cooperating is best for both the individual and the group.

We can contrast this perspective against arguments that community comes first and individuals second, arguments that sacrifice individuals and subcommunities for the larger community, whether it be limiting the opportunities of individuals within that community, demanding work or resources from individuals, or even enslaving or killing individuals. It is one thing to sacrifice yourself to the cause and another thing to sacrifice another person. It is one thing to ask everyone to find their role within the community and another thing to force people into roles they may not want. Importantly, a community that requires a scapegoat is not solving the assurance game. Neither the winner (loser) in Shirley Jackson's *Lottery* nor the child tortured in the basement in Ursula K. Le Guin's *The Ones Who Walk Away from Omelas* is doing individually better in their corner of the game being played.

As a sharp contrast between the morality argued for within this book and collectivism, we can look at the moral parable of the one person standing up for inclusivity, equality, and fairness in the midst of a mob. We celebrate the lone voice of reason in a mob or the small group dedicated to changing the views of a nation. Like all moral myths, these stories are about first-person aspirational goals. They are about trying to change the minds of the group into a more inclusive point of view. Notice that these myths never run the other way. These myths never celebrate the people who stay and give in to the mob nor do they celebrate the one person shouting invectives at the passersby in the street.

The Sociobiology and Naturalistic Fallacies

One common assumption is that the world is getting better, and thus whatever moral structures we observe are the answers we seek. While there is pretty good evidence that the world really is getting better given our assurance game goals and overall individual welfare,[3] that doesn't mean we're done or that the observed moral codes are the best available. Concluding that the observed moral codes are the best is an example of the *sociobiology fallacy*, which assumes that observed sociological structures (such as societies) derive from an optimization process, and the *naturalistic fallacy*, which is the idea that those things that occur are good. These fallacies make two assumptions that are wrong.

First, they assume that the optimization process (usually attributed to genetic evolution but sometimes attributed to competition between societies) has completed. But these evolution-like processes are processes; they are slow, have long historical dependencies, and chase a changing landscape. Genetic evolution is not optimizing for the most efficient animal or the "best" animal overall; it is optimizing for the traits that match the current situation the subject faces. As that situation changes, what is "best" can change dramatically, even on fast timescales (say, from year to year). (This is one of the problems being created by climate change—the environments many species are finding themselves in are changing quickly, and traits evolved for one environment are not as useful in another.[4])

Second, the idea that we can measure morality by the observations of current societies assumes that the goal of morality is to create stable societies. It leaves us with a nihilistic view of humanity and the concept that might makes right, that the winners write the history books, and that whatever societies exist today are the ones to emulate.

In practice, many attempts at scientific descriptions of morality end up with the statement that one can explain the neurophysiological and sociological mechanisms underlying tribalism, genocide, and caste construction, but one cannot identify whether any of those mechanisms are better than others. (This is the *parochial altruism* that we keep coming back to as a description of humanity.[5]) Other scientific descriptions fall back on basic ideas of utilitarianism for the prescription but admit that no human society follows that utilitarianism. They note that the prescriptions of utilitarianism sound insane when stated bluntly and so fall back on a prescriptionless view of those neurophysiological and sociological mechanisms.[6] Both of these perspectives lead to an amoral view of the description of human behavior in relation to moral questions.

Haidt's Moral Matrix

One of the most recently influential takes on morality via descriptive observation of human statements and behavior is that of Jonathan Haidt's five "tastes" that he calls *moral foundations theory* and that he popularized in his book *The Righteous Mind*. Haidt argues that measures of morality across the world can be categorized into five realms in which humans make moral decisions: care (harm), fairness (cheating), loyalty (betrayal), authority (subversion), and sanctity (degradation). Notice that each of these perspectives reflects questions of the assurance game (supporting or disrupting it). Care (harm) and fairness (cheating) directly address questions of cooperation and helping in need. Loyalty (betrayal) is about asabiya, the willingness to work within the group. Authority (subversion) is about the willingness to respect the structure of the group and the rules themselves. Sanctity (degradation) accesses the Pavlovian processes of awe and respect (on the positive side) and disgust and repulsion (on the negative) that have been conditioned by culture to create tools to drive other behaviors within the group.[7]

While we can see these "tastes" as tools to change behavior and Haidt does argue that these are important steps to evolving social dynamics, he also treats these as moral goals in themselves. Haidt explicitly refuses to say whether the different choices made by different cultures with regard to these goals are good or bad—only that moral questions address these five issues. In doing so he falls into the error of the naturalistic fallacy. For example, in an attempt to reduce his self-perceived "Western, individualistic bias" Haidt looks at societies in India in which women are limited in their opportunities and sees those societies as good within the "morality of community" and the "stability of family."[8] I suspect that Haidt is empathizing with the men in this situation more than the women, and I wonder if he might have come to a different conclusion had he empathized with the women with limited opportunities. Even if these limitation constructions are enforced by people in the oppressed/limited subgroup (as they often are), that doesn't change the fact that these limitations push games away from true assurance games into those in which individuals are sacrificing (or are forced to sacrifice) their share of the benefits. Our statement that the goal of morality is to increase the assurance game says that moral structures that provide for both parties are better than moral structures that limit one party (limitations push the limited party into the *I cooperate while you defect* corner of a game that is very much not an assurance game).

An important issue in any question of morality is *Who is included in the protected class?* A naive view of the assurance game produces parochial altruism, cooperation within the group and conflict (xenophobia) beyond it. An important factor that we will address in the second half of the book is defining which individuals are part of that group. We will see systems that define groups genetically, such as through family lineages; systems that define groups through superficial characteristics, some of which are more easily modifiable (language and accent) than others (skin color); and systems that define groups politically, through national boundaries, shared religions, or shared beliefs, as well as many other systems of social construction. Many of these groups have historically defined themselves in contrast to a negative "other." However, it is important to remember that these group definitions are not the ultimate goal that our moral structures are aiming toward. The actual goal is to generate a cooperative assurance game because cooperation produces positive non-zero-sum benefits. Creating enemies is only one tool in the moral arsenal. Other tools might work better.

Tools to Improve Society

In his book *The Better Angels of Our Nature*, Steven Pinker documents the astounding decrease in violence over the millennia and identifies five tools that have been implicated in that decrease. Pinker argues that these tools have changed both the payoff matrices of the prisoner's dilemma and how our neuropsychology interacts with the world.[9]

First, there is *Leviathan*, the state that imposes external penalties on defection, making the negative choices less valuable. The second is *commerce*, in which additional opportunities arise from cooperation, making the positive choices more valuable. The third is what Pinker calls "feminization," which is really about changes in the moral codes that shift *glory, honor,* and *societal rewards* from violence to nonviolence, providing fewer rewards for violence and thus shifting gains from the defection to cooperation choices. The fourth is the *expanding circle,* in which perspective-taking and empathy makes one appreciate the negative consequences of violence on victims. And fifth is the *escalator of reason,* in which logic is applied to aspirational goals, finding inconsistencies and increasing universality. In addition, Pinker notes concomitant increases in empathy and self-control over the millennia, largely due to sociological processes that have increased our knowledge of (and thus empathy for) others and trained our self-control abilities. Pinker also points out a number of smaller etiquettes and moral codes with indirect consequences, such as the way sharp knives have been removed from most dinner tables in many cultures.

We are addressing moral tools beyond their implications for violence, although the reduction in violence is, of course, an important component to that developing morality. Pinker's explanation for the reduction in violence as due to sociological tool development is a great example of how moral codes change behavior, partially through changes in those payoff matrices and partially through their interactions with our neurophysiological decision-making systems.

Importantly, we will see that these are only some of the social constructions available to us in our moral design and that some of these social constructions work better than others, in part because they are more likely to change behavior, in part because they include more people in the group, and in part because they allow those people to better reach their individual potential. Because the assurance game is non-zero-sum, as the in-group grows to include more and more people and as more people reach their individual potential, not only does society do better; individuals within that society do better. To come back to the epigraph at the start of the book, "We all do better when we all do better."

Descriptions of how humans interact with moral questions cannot be seen as the goals of morality but must be seen as tools that are available to address questions of social structure. We can ask how each of these tools can play a role in helping to construct and maintain society. We can explore these and other social technologies and ask how these tools interact with our decision-making systems to change our behavior. The descriptive science of how humans behave (the *is* question of morality) cannot provide us with the goal of what humans are trying to achieve in the moral realm (the *ought* question of morality). Instead, we have to turn elsewhere to understand the goals that underlie the decisions to use those tools. In the first part of the book, I made the case that the goal is to find our way to an assurance game in which others will cooperate with us, creating worlds where we can reap those non-zero-sum benefits that arise from working together.

ONWARD

We can think of the first half of the book as asking "scientific questions" of morality; the second half of the book will ask "engineering questions" of morality. Science questions and engineering questions are different. Science questions are about how things work in the world: *How do humans react to moral questions such as the trolley problem? How do optimal decisions change between the prisoner's dilemma and the assurance game? What are the consequences of adding third-party punishment to a tax game?* Engineering questions are about how to build a better structure. For example, given the importance of empathy in moral questions, we can ask about ways to increase empathy (desegregation, increasing conversations between groups).[10] We can also identify that some solutions, such as building walls, decrease empathy, which makes it harder to empathize with the other group and produces increased anxiety and mental discomfort in people living near the wall and in both groups on each side of the wall.[11] Because we have identified normative goals in morality, we can say that some of these engineering solutions are more moral than others.

This point that we can use science for good or ill is not unique to morality. We can use a knowledge of physics to build a rocket to the moon or to sabotage that rocket and ensure that it won't fly. We can use a knowledge of how human bodies work to cure disease or to poison someone. We can use a knowledge of mental health to help an individual or to break them. But, as we've noted, these sciences also have normative goals. I think people would have been quite upset if someone had sabotaged the *Curiosity Rover* and prevented it from reaching Mars. Doctors swear an oath to "First, do no harm." The American Medical Association and the American Psychiatric Association refuse to allow their members to participate in interrogations, particularly "enhanced interrogations" commonly known as torture. While the American Psychological Association initially agreed to let their members participate, they have since rescinded that stance.[12]

While we have found some interesting answers to some of these questions, the science is not finished, nor is the engineering. The discoveries we've made open up new questions to ask (as any good science should do). The sections in the second half of the book can provide some guidance forward, but they are not a recipe for how to create a better society. They are the kinds of experimental questions that we will need to ask about morality.

We've primarily explored the unwritten rules that underlie human society so far and have noted that they tend to be based on the intrinsic goals required to form a society (encouraging identified roles within a group, reducing intragroup strife, and providing for third-party punishment while allowing a process of redemption). Next we will explore the codified structures we have created, including legal structures, religious structures, and governmental structures. We will find that these structures create a multitude of levels that interact, and while competition seems to be a necessary component to create group-level selection (it is the competition between groups that makes a group of cooperators more successful than a group of selfish players), nothing in the mathematics we've seen says that the competition must be violent or even that it must entail conflict (sports teams compete but are not in conflict).

What is interesting from a *science of morality* perspective is that we can study these interactions both from the aspirational side of constructing a community of non-zero-sum benefits and from the actual side of the interaction that individuals experience. By thinking about communities, about cooperation and defection, and, most importantly, about how our knowledge of neuroscience affects our actions, we will find a new view on how the aspirations of our humanity suggest new social structures that will better achieve those goals.

This is only the beginning. There is a lot of research to be done, but I hope readers can see that this new view on morality as social technologies interacting with decision-making systems to change the rules of the game puts these attempts in a new light.

* * *

CAVEAT

I feel comfortable claiming expertise in both the neuroeconomic and decision-making literatures discussed in the first half of this book; however, I am definitely not an expert in the literatures discussed in this second half. These next chapters depend on questions of sociology, legal thought and scholarship, religious studies, governmental design, macroeconomics, environmental ecological economics, and moral ethics, none of which are fields I have explicitly studied, and I feel woefully unready to take on these large and important topics. However, this book would be incomplete without a discussion of them or their relationship to the ideas laid out in the first part of the book. I have done my best to understand the literature, to talk to colleagues working in these fields, and to bring in what information I could find, but readers should recognize that these next chapters are going to be incomplete. Instead of taking these next chapters as definitive reviews, I hope that my readers will take them as first thoughts on the consequences of the neuroeconomic and decision-making literatures on moral design. And I hope that my colleagues in these fields will take them as an invitation to collaborate, as a letter put into a bottle and placed in the scientific ocean with the hope that someone will find it, see the connection, and reach out to collaborate. Because science itself is an assurance game, and the work we could do together is far greater than the work we do alone.

III INSTITUTIONS AND POLICIES

13 THE RULES OF THE ROAD

If we change the rules of the game, we change the decisions that get made, and we can make the world a better or a worse place.

A society is an organized community, a group of people with a set of (possibly unwritten) rules and social norms that allow them to work and live together. Some of those rules are arbitrary, but we need them for our society to function smoothly. It doesn't matter which side of the road we drive on as long as we're all driving on the same side of the road. It doesn't matter what we use for money as long as we all agree to accept it as a medium of exchange. In fact, we don't even need to have a physical thing (a bit of metal, a piece of paper) to hand between each other as long as we all agree to accept the identification of the numbers so that I can trade you the right number of monetary units for the thing I want to buy from you. Famously, on the island of Yap, what was traded as money was the community-defined ownership of certain large (unmovable) stones. Most transactions today are handled digitally and never translate into a physical exchange of monetary objects.

Other rules, however, are attempts to level the playing field, to make the game we are playing more of an assurance game, and to increase the likelihood of cooperation within our society. Or to unlevel it. One of the important points we will see in our engineering discussion is how rules can also be used to empower one person or group above another. In a very real sense, rules used as a tool for power are changing the game from an assurance game into something else, leaving the oppressed party with only bad choices.

Specifying exactly what this other game is would require working out the details of the options. How should we describe the choices of an enslaved person when the available options are giving your work away versus being whipped or killed if you don't? One of the points of this book is that because the assurance game is non-zero-sum, these negative power changes are (in the long run) weaker than getting both parties to play an assurance game together. One of the reasons the Union won the American Civil War was because the enslaved population left the plantations to work for the Union army and even to join it and fight back directly. We can see this as a situation in which the enslaved population had an opportunity to defect against the Southern slavers who were defecting against them because the Union was providing more of an opportunity to cooperate.[1]

Rules can even create playing fields that didn't exist before. For example, stock and trading markets only exist under rules and regulations. An open-ended interaction between two people or two groups of people (that can, for example, have decisions based on physical bullying or other violence) is not a market. In the same way that a sport with rules is very different from a no-holds-barred war, a well-regulated market is very different from an open-ended interaction. Of course, as we've seen, even on an unnegotiated battlefield, soldiers encountering other soldiers tend to "posture" and "submit" rather than immediately kill each other.[2] Most interactions between people contain underlying regulations due to our Pavlovian interactions, which arise from evolutionarily built systems that also exist in nonhuman primates and other animals.[3]

These new rules can create new assurance games to play and can have consequences beyond their immediate changes. For example, one of the most successful environmental regulations has been the creation of a cap-and-trade market for environmental pollution.[4] Pollution is a classic example of a problem of the commons seen as a generalized prisoner's dilemma. Reducing pollution costs the individual, but the consequences of not reducing the pollution are spread out among the community. If the person or company releasing the pollution into the environment can release it into a place they are not, then it hurts them even less. Environmentalists were initially skeptical of cap-and-trade as a mechanism to reduce pollution because it gave polluters the right to pollute. But then environmentalists found an interesting effect of the cap-and-trade rules—they could buy up the right to pollute and hold on to those pollution rights (without actually using them). This trick shrinks the number of pollution rights that exist and drives up the price of the pollution rights. Once the pollution rights become more expensive than the cost of reducing the pollution for some given factory, it becomes economically sensible for the person making the decision to reduce the factory's pollution.

Cap-and-trade created a new game, a market for pollution, and in doing so made it worthwhile for an individual to reduce their own pollution. Sure, one could have simply said, "The factory shouldn't pollute," but that's asking them to be moral because it's good for the group despite the fact that it's bad for them as an individual. One could have simply provided third-party punishment (fines) for polluting at all, but that requires a large amount of oversight and encourages defection as a means of getting around those regulations. Instead, the cap-and-trade market changed the rules so that (once the price of these pollution rights had gone up) it was actually better for both the factory and the community to reduce the factory's pollution. Cap-and-trade changed pollution from a prisoner's dilemma (better for the community to reduce pollution, better for the individual to let the factory pollute) into an assurance game (better for both the individual and the community to reduce the pollution that the factory was spewing out).

INSURANCE

One of the major roles of community is to provide support in times of trouble. A classic economic description of this is the question of insurance. A disaster can be extremely expensive. Even if we start from the simplifying assumption that an individual could save enough money over a lifetime to pay for that disaster, there is no reason to think the disaster is going to occur at the end of their life, after they've saved up enough money. What if the disaster happens at the beginning? Instead, the obvious answer is to amortize the cost over the community. If each person pays a small amount into the central pot every year, we can be sure that each person is covered should that disaster strike. We can turn the risk of an economically debilitating disaster into a small, known cost. This is straightforward economics. It is also, exactly, the tax game that we looked at in chapter 3.

This leads to the question of taxes and a vigorous, ongoing debate in the US right now about whether people should be working together or alone. A fascinating example that shows the moral quandary within these tax games was a Tennessee house fire in Obion County in 2010.[5] The local county did not provide unified fire protection, but the nearby town of South Fulton had a fire department willing to respond as long as the homeowner paid a $75 annual fee. The fee was not mandatory (it was not a tax), and thus some owners paid and some did not. Other nearby counties with a similar inability to tax residents sent a bill after services were rendered, but many homeowners failed to pay the bill after the fact, possibly because they didn't feel they needed to, the service being completed, and possibly because the cost was expensive and they lacked the funds to cover it. In the US today, almost half of Americans do not have enough savings to cover a $400 emergency expense.[6]

Presumably, as with any insurance, the $75 was the amortized cost for South Fulton's fire department to come over and put out the fire. However, the homeowners had not paid the fee. Because their neighbor had, the fire department showed up and watched the first house burn, putting out the fire when it jumped the yard to the neighbor. I find this example a fascinating moral question because the homeowner, of course, offered to pay the fee immediately (but $75 is not the cost of the fire truck's run). News reports said that the owners and their neighbors offered to pay whatever was necessary to put the fire out, but the fire department refused. It certainly would not have made sense to let the owner pay the $75 fee. On the other hand, not only did the fire destroy all of the family's possessions, but the family pets died in the fire, including three dogs and a cat. Would the firefighters have saved a child or other family member had they been trapped inside? The response to the Tennessee fire incident was very mixed. The firefighters did say that they would have intervened if a person had been trapped inside, but that can be hard to determine, and one does not want to disregard the emotional devastation of losing the family pets who died in the fire.

There has been a strong push to reduce the size of government and privatize all governmental services over the last forty years, which means that one has to pay for any help one gets, but this leaves open the question of what to do in emergencies. The Tennessee fire incident brought the cost of not paying that $75 fee to light. Subsequent to this incident, the South Fulton fire department now charges $3500 for the response if the $75 fee has not been paid. Of course, that raises the question of what the fire department will do if the owner doesn't have the money available and what the fallout will be when the homeowner gets the bill. Studies of the situation found that the cost to taxpayers would have been $42 per household per year if the cost had been assessed as a tax that everyone paid rather than a subscription fee.

This is the tax game, plain and simple.

As I write this in 2021, the question of medical treatment and its inherent costs is a key subject of debate across the US. I am writing this in the midst of the continued COVID-19 pandemic, during which a surprisingly large portion of the US population has refused to be vaccinated, and the latest COVID-19 surge running amok among the unvaccinated is filling the hospitals and emergency rooms, making them unavailable for other needed treatments (such as heart attacks or injuries from car crashes). It appears that the unwillingness to take the vaccine is occurring as a *tribal marker*, a way of defining subgroups. Our identification of the goal of morality as being to increase cooperation within the assurance game (which will make the world better for both parties) implies that this tribal marker is a particularly poor moral code. In fact, vaccination is not even an assurance game. Since the COVID-19 vaccines are safe and effective, it is better to take the vaccine even if the rest of the population doesn't. Refusing the vaccine as a way of marking membership within a tribe is a poor moral code that has a negative impact not only on the people beyond that subgroup but on the people within the subgroup as well.

But this debate predates the COVID-19 crisis because it is illegal for emergency rooms to turn away patients who need medical treatment.[7] (Note the obvious analogy to the moral statements by some that the firefighters should have put out the house fire, even though the owner had not paid the fee.) However, while it is illegal to turn a patient away from the emergency room, there are no federal support funds to pay the cost of that emergency room visit. Hospitals have to recoup those uncovered emergency room costs somehow, so they increase the costs to other hospital users, effectively forcing those other users to pay a tax. (Nationalized health care similar to what exists in most other countries would solve this problem.)

But more than that, this law means that emergency rooms are available as a last resort, even for people without health insurance, which hides the fact that some people with limited or no insurance do not receive preventive care. (Preventative care is generally much cheaper than addressing a problem only after it festers and becomes dangerous, which means that our national health care costs, as a society, are larger than they need to be. We are all paying the cost for this incomplete preventative care coverage.) The current discussion over the insane costs of US visits to the hospital emergency room (thousands of dollars for a bandage or a blood test that normally cost pennies or a few dollars) has cast the hospitals as evil, when the real problem is that we have asked the hospitals to amortize the costs themselves instead of handling the amortized costs as a society.

FAIRNESS

One of the most interesting things about morality is that morality is aspirational. From chapter 9, we can see that this is because intrinsic goals have evolved into humans through the evolutionary pressures of living in a social group. This means that we would like to live in a moral world—one in which we find ourselves in a cooperating group that wins the assurance game. Unfortunately, these goals are not always as easily achieved as we might like. While some have called humans the "moral animal" and we have made a case for that argument throughout this book, studies of human behavior generally find that people show parochial altruism, cooperation within a group but xenophobia outside of it. This book is not about that parochial altruism. It is about how our understanding of the mechanisms of the assurance game and how our understanding of the decision-making systems allow us to engineer social structures (moral codes) that can increase the likelihood that we are participating in an assurance game, that we are playing fair with the community. This book is about how changing the rules of the game within a group changes both who can be in the group and how the group works together.

What is fascinating is that our societies, in both extensive studies of small tribal societies of tens to hundreds of people and the founding documents of large civilizations of millions of people (such as the 1776 US Declaration of Independence, the 1789 French *Déclaration des droit de l'homme et du citoyen* (Declaration of the Rights of Man and of the Citizen), or the United Nation's 1948 Universal Declaration of Human Rights) speak aspirationally of fairness, that all people are equal before the law. Of course, one must be careful about equality of opportunity and equality of outcome. As the French novelist Anatole France famously pointed out, "The law, in its majestic equality, forbids both the rich and the poor alike from sleeping under bridges, begging in the streets, or stealing a loaf of bread."[8] If the law were instead to "forbid both the rich and the poor from going hungry," we would have a very different society. (These are different moral codes.)

The Declaration of Independence, one of the key aspirational documents of the United States, starts by stating "We hold these truths to be self-evident, that all men are created equal." This document specifically spoke to a gendered difference (men, not humanity), included a society in which all men were certainly not seen as equal, and was written by a man who owned slaves and even sired children with one of them.[9] Nevertheless, this document has been interpreted to include larger and larger groups over the subsequent centuries. Modern interpretations of this aspiration have removed the gendered view and the racial view, and chattel slavery is completely gone from our modern world.

Of course, no one looking at any modern society would argue that we have achieved gender or racial parity across all aspects of our society. While chattel slavery (permanent slavery based on the "ownership" of a person specifically, including all of their children) may be gone, other forms of forced work, including wage slavery, slavery through physical or legal threat, and slavery through penal punishment, still exist. Their reduction and elimination (including the important gender and racial disparities that exist) are still a work in progress in the US and many other modern societies. Nevertheless, we can talk about fairness and the aspirational goals of fairness.

One of the clearest aspirational goals in the philosophy of fairness is the point made by the philosopher John Rawls, who noted that it would be logically acceptable to have inequalities in a society if everyone agreed to those inequalities before knowing where one would fall within that inequality. That is, one has to agree to the societal structure before one finds out where one lands within it. Notice that this logic does not imply a society of pure equals—just that the probabilities of where one ends up in the society are equal. This Rawlsian structure underlies a lot of human morality. I think, for example, that this Rawlsian perspective is the key to understanding the Milgram experiment we talked about in chapter 8. The subjects of the experiment thought they could end up as either the teacher or the learner and thus were willing to punish the learner because they would have (I suspect) willingly accepted punishment had they ended up in that role.

I don't want to be a Pollyanna about this. This Rawlsian perspective is aspirational. I suspect that most people do not think they could have ended up anywhere within society, and many people, particularly aristocrats at the top of a hierarchy, think they were destined for that position within the society. People expect some inequality within their society, particularly modern societies. Importantly, however, people vastly underestimate the actual inequality within societies.[10]

Let us take the question of the aspirational goal "Everyone is equal before the law." Of course, this is not true in practice. We all know that more powerful people are not punished at the same rate as less powerful people and that there are racial, gender, and social disparities in how people are treated by the law.

We know that there are very different interactive experiences with police and the legal system as a function of socioeconomic class, as well as racial and gender identity. Ta-Nehisi Coates, in his book *Between the World and Me*, talks eloquently about the dangers of being a Black man in the US today and about the practicality that his son would need to prepare for in order to deal with the unfairness in the world. I am always amazed that the police took James Holmes (a White man who killed twelve people and injured seventy others in Aurora, Colorado, in 2012) alive but killed George Floyd (a Black man accused of passing a counterfeit $20 bill). In the civil protests and unrest in 2020, we were shown a side-by-side comparison of the police reaction to both anti-mask and anti–racial murder protests. The anti-mask protests consisted of right-wing, mostly White agitators armed with loaded semiautomatic rifles who screamed in the faces of the police that we should open up the economy in the midst of a pandemic (and that somehow wearing a cloth mask over one's face was a crime against freedom?). The police stood quietly in cloth uniforms with simple cloth face masks. The anti–racial murder protests consisted of unarmed, racially diverse protesters kneeling six feet apart (socially distanced to prevent spreading COVID-19). The police came at them in full military riot gear. This shocking difference has become an icon of this inequality of treatment before the law.

It is an important societal question to ask ourselves where the police fit into our moral story. Theoretically, police should serve as a community self-control (see chapter 6), giving society an opportunity to provide third-party (graded!) punishment. (Importantly, theoretically, the police are not supposed to provide the punishment themselves.) However, the history of policing in the US is fraught with complexity. On the one hand, it has included a long history of community service and support in times of emergencies, while on the other, it has also included a long history of policing as an oppressive force maintaining an inherent caste and power structure.[11] One of the complex issues today is that as government support for services has been eviscerated, the police have taken on larger and larger community service roles for which they are poorly trained and which their techniques are poorly suited for. It has been very difficult to punish police officers who take the militarized steps of an occupying force, in part because many police unions see themselves as separate from the communities they are supposedly protecting. There is a movement in the US right now to shift the police from a militarized warrior mentality to a community service mentality (a shift in moral codes). How that will play out will be interesting to watch over the next few years.

I am not going to defend our current legal systems as ideal. We know that they are not. But understanding the goals of legal systems as a means of "changing the rules of the game" will allow us to identify good rules for good legal systems in order to better our societies and enable each person to find themselves in the cooperate-cooperate corner of the assurance game.

A very interesting example of how societal perspectives interact with decision processes to change behavior and its consequences can be seen in the history of how society has viewed addiction. Most current views on addiction in the scientific community are based on the idea that it is a dysfunction of the decision-making processes and their interaction with the social and other environmental situations that these people encounter.[12] Importantly, current scientific views on addiction separate it from drug use per se. Current scientific interpretations include behavioral addictions that do not arise from drug use, and drug use that does not include an addictive dysfunction.[13] That being said, there is often a disconnect between the scientific view on addiction and the political rhetoric used in discussing addiction, which can make implementing the right treatment difficult.

This view of addiction as a dysfunction in decision-making has an important history because addiction was once seen as a moral failing.[14] It was seen as an inability of willpower, and people with addictions were cajoled to simply "shape up" and "behave better" using third-party punishment, religion, and other limited moral tools available at the time. While this may have worked for some individuals, in practice, it did not work for most people and did little to prevent addiction.

It is interesting to see addiction in the light of Samuel Butler's alternate-reality novel *Erewhon* ("erewhon" is, of course, "nowhere" spelled backwards . . . or at least it almost is), in which crime is treated as a medicalized problem that requires sympathy and treatment, while illness is seen as a societal problem that requires punishment. In the land of Erewhon, those who are sick are punished with fines and imprisonment, while criminals are provided medicines and sympathy and "inquiries as to when the symptoms first started." *Erewhon* was written as Victorian satire, but we will see that the interaction between the medicalization and punishment of both disease and criminality is far more complex than usually described. Below, we will see how our view of addiction was transformed from criminality (seen as a personal moral failure) to a disease to a symptom of an underlying dysfunctional interaction between the brain and the environment that can be treated both neurophysiologically and environmentally (socially). If we follow this history carefully, we can see how we can use our understanding of human decision-making processes to change societal structures that can improve addiction treatment.

A History of Society's Views on Addiction

In the late nineteenth and early parts of the twentieth century, addiction was seen as a problem of willpower, treated as a social failing, and punished with ostracism and disdain. Interestingly, these early views were based on concepts of temptation. While some legal structures were put in place to criminalize drug takers, most of the major steps taken to prevent drug use were aimed at reducing temptation. China restricted opium imports (until the British forced them to import it at gunpoint[15]). The Eighteenth Amendment to the US Constitution stated that it would be illegal to manufacture, sell, or transport "intoxicating liquors," not that it would be illegal to drink those "intoxicating liquors." The problem, of course, is that people with addictions are highly motivated to achieve those addictions and will pay high costs, take on extreme dangers, and risk interacting with criminal enterprises for those addictions.

In the 1960s the scientific community began to see addiction as a disease and a brain disorder and began to medicalize it. People with an addiction were seen as having a medical problem that could be treated with pharmacology or other medical treatments.[16]

At this same time, however, the US government began a "war on drugs" that created a military response to drug use. Interestingly, both the original Nixonian speech that defined the war on drugs and the major official push during the Reagan administration ("Just say no") proposed education-based rather than military steps,[17] but they used military language and began an increased militarization that criminalized drug users. Punishment was enforced for drug possession, not just the creation of temptation. The jailing of drug users increased even as it became clearer that drug treatment produced better outcomes. This has produced dramatically negative effects throughout the US. Note that this is the science of morality—we can identify clear consequences of our choices. Jailing drug users produces worse outcomes for society, including increased problems of relapse, increased probabilities of transitions to violence post incarceration, and decreased participation in society during and after incarceration. It is, simply put, far more expensive to the taxpayers to jail drug users than to send them to drug treatment facilities.[18] These descriptive consequences mean that the two prescriptive choices (jail, treatment) are not morally equal.

Nevertheless, by this point in our history (the 1970s), scientific views on addiction had moved from a "choice" and a "lack of willpower" to a "brain disorder."[19] A number of recent popular science arguments have been made that this "brain disorder" language misses the fact that many of the reasons for drug addiction are due to interactions between the natural learning systems in the brain and environmental situations. However, it is important to recognize that this "addiction is a brain disease" language has always included an (often unstated) psychosocial (neurosocial) interaction between neural function and the environment. Addiction has long been talked of as experiences "hijacking" the brain's natural learning systems. The experiences of the addictive behavior (the chemistry of the drug in question, the timing of the gambling wins and losses, the sexual situations) interact with the brain's natural learning systems, which drives the decision-making systems into problematic computations. Currently, addiction is seen as a disease (disorder) of choice—as a dysfunction in the mechanisms by which we make decisions.[20]

Obviously, many of these disorders have societal consequences. One of the most profound effects of addiction is on one's family and friends. As the addict finds themself sinking deeper and deeper into the hole of their addiction, their relationships suffer dramatically. Although addictive behaviors can occur in highly social situations, many people describe addiction as deeply lonely.

In fact, some individuals have found paths out of their addiction through sociological interactions. Gene Heyman, in his book *Addiction: A Disorder of Choice* (meaning a disorder of the way we make choices, not that addiction itself is a choice) points out that many individuals finally decide to go to treatment when their addiction interferes with people they love (like the mother who decided to get help and treatment when she realized her preteen daughter was starting to follow in her footsteps). Many individuals have found their way out of addiction by finding new social hobbies with groups that can provide alternate choices (rock climbing, running, book clubs, or church socials).

Addiction Policy

From the neuroscience of addiction, we can identify four policies to help people with addiction.

First, we need to recognize that addiction is biological. That is, because we are physical beings, people with addictions often have physiological changes that need to be addressed. This is the point of nicotine replacement therapy (such as the nicotine patch) or methadone treatment for opioid addiction.[21] But, often, these physiological treatments are not enough because addictions, in large part, result from the misappropriation of our natural learning systems, and even after the physiological and pharmacological dysfunctions have been restored to their normal function, the mislearned associations remain.

Second, we can increase the cost of the addiction. Economically, we can measure how quickly a subject becomes unwilling to pay for a thing as the cost increases. If the baseball tickets just doubled in price, are you half as likely to be willing to go to the game? A thing that drops off quickly is *elastic* (typical of luxury goods such as jewelry or sports tickets), while a thing that drops off slowly is *inelastic* (typical of necessary goods such as food). While it is true that addictions do show some elasticity to costs—that is, people decrease drug use as costs increase—in general people with addictions are particularly inelastic with their drug-taking decisions.[22] Some people have argued that this is the definition of addiction: addiction is the inelasticity to costs.[23] So punishment might not be so particularly effective.

Third, we can limit the temptations available by limiting the opportunities to achieve the addiction goal. This is, of course, the point of anti-drug laws like the anti-alcohol temperance of Prohibition and the Chinese anti-opium laws of the late nineteenth century. The problem is that addicted drug users are willing to pay high costs for their drugs, which drives them to search out even dangerous sources. Laws restricting availability can even increase the criminal element of drug suppliers.

Finally, we can teach people self-regulation strategies. We can provide societal structures that make it easier for them to apply that self-control and avoid their addictions. One of the most effective treatments turns out to be providing rewards for avoiding addictions in a treatment protocol called *contingency management*.

Self-Regulation

The key insight from the theory that our behaviors arise from multiple decision systems (chapter 7) is that we can be inconsistent. While one decision system may want to take one action, another may want to take a different action. This is the main idea underlying concepts of *self-control* or *self-regulation*. These terms imply that we are not unitary decision-makers—there must be a self that is controlling (regulating) and a self that is being controlled (regulated).

I prefer the term *self-regulation* to that of *self-control* because self-regulation suggests that one can use a host of techniques to help one avoid things one would like to avoid, whereas self-control suggests that it all depends on willpower. Because willpower takes effort and effort is impossible to sustain forever, depending on willpower to avoid temptation will eventually lead to failure. But it is possible to develop good habits that lead one away from temptation and to choose to precommit to decisions that avoid temptation before one is actually tempted. An example of a good habit is changing one's drive home so that one never passes the casino. An example of precommitment is putting money in a certificate of deposit so that it cannot be retrieved until a certain time in the future and therefore can't be spent gambling at the casino or on the stock market. Another example is pouring all the alcohol down the drain so one cannot drink it.

Other ways of improving self-regulation include *bundling* and training oneself to attend to future consequences (*episodic future thinking*). In bundling, one changes what one thinks the outcomes are by combining future decisions. Thus, if one can change the question from "Should I have this one drink?" to "If I drink, I'll drink a lot, so the real question is do I have a lot of drinks or do I not drink at all?," one might come up with different decisions. Someone who thinks there is "no such thing as one drink for an alcoholic" may be more likely to decide to not drink at all.

Many people have talked about addiction as a problem with attention being too focused on present gains over future gains. People in general attend to the present over the future. One reason for this is that the future is uncertain and abstract. As seen in chapter 7, deliberation and planning depend on actually imagining a future outcome. If we make the future more concrete, it becomes easier to imagine, and people are more likely to choose future rewards over present temptations. It is possible to train people to get better at imagining those future outcomes. That training decreases drug use.[24] By understanding the computations underlying decision-making, we can design treatments that change behavior.

A good way to think about these self-regulation concepts is that one of our multiple selves can change the rules of the game to make it easier to behave in a certain way. Just as we've seen throughout the first part of the book, we can change the game to make it harder to take an action we don't want taken (pouring the alcohol down the drain, making us go out to get it) or easier to take an action we want to take (having a friend or "sponsor" to call anytime day or night to talk us out of going to the bar and drinking).

Contingency Management

The most important consequence of the multiple-decision-maker theory is that *how you ask the question changes how the person makes their decision.* This means that we can create institutions that change behaviors by forcing changes to the question with policies that nudge people in the right direction.

One of my favorite examples, particularly in the context of addiction, is contingency management, in which we basically pay people to not engage in their addictive behaviors (such as taking drugs).[25] In contingency management applied to addictive drug use, a recovering addict proves to a clinic that they have not been using drugs for the previous time period (usually a week) and are given a small reward (a few dollars, a free movie, a few points toward a large purchase, a chance to pull a prize from a fishbowl—the prizes can range from a bottle of shampoo to a new iPad, but the average reward from these prize versions remains small). The usual description is that contingency management creates an economic cost to using drugs—using the drug not only costs whatever the drug itself costs but now also costs you the loss of that extra reward.

The problem is that drugs on the street are notoriously economically inelastic. People have done studies and found that while increases in cost do reduce drug use, those effects are small. If the only effect driving the success of contingency management was the increased cost of losing that reward, contingency management would have little effect on people's addictive behaviors. Simply put, contingency management works much better than it should.[26]

An alternate explanation for the success of contingency management is that it shifts the decision from an *Is it worth it?* decision to a *choose-between* decision. Instead of saying, "Is it worth this much money for my drugs?," the user can now say, "Would I rather have my drugs or that gold star?" and can say, "I can wait for that gold star." Effectively, contingency management may be shifting the decision from a Pavlovian- or procedural-driven decision into a deliberative-driven decision.[27]

CHANGING THE RULES OF THE GAME

Contingency management is a way for society to change the rules of the game. Like Nick declaring to Ibrahim that he was going to steal the money in their split-or-steal interaction (chapter 2), like the cattle ranchers in Northern California or the lobstermen in Maine with a set of retributive rules they all know and expect (chapter 6), we have changed the game people with addiction are playing and have made it easier for them to change their behavior.

Contingency management is a form of a *nudge*—it helps change behavior but doesn't force that behavior on anyone. Understanding the neuroscience of decision-making provides us the opportunity to create changes in policy that steer people toward directions we would, as a society, like them to go without forcing them into that choice.

A classic nudge is that of the default option. Because changing from a default option to another option takes cognitive effort (one has to think through the consequences), many people who don't care very much or don't want to take on that cognitive effort simply accept the default option. Generally, the decision process entails first asking, "Is the default option good-enough?" If it is, leave it. If it is not, only then spend the cognitive effort deciding what to choose. Two great examples of default option nudges are setting default retirement plans and checking the organ donation checkbox, both of which have societal (and moral) consequences.[28]

Retirement plans are complex things. They include a great deal of math that depends on expectations of one's earning potential over a long future time course, expectations of one's expenses after retirement, and expectations of how different economic products will gain or lose money over a complex future. Many people given choices of retirement plans simply throw up their hands and give up. When people have to sign up for retirement plans or they get nothing, a surprising number of people simply don't sign up at all—even though any option is better than none. By providing a pretty good plan that has been worked out by experts as a default, most people will take that plan. While it might not be the absolutely optimal choice for them as an individual, it guarantees that people have some sort of retirement plan. Of course, one can also provide the option of opting out of any retirement plan, but by making that an active choice, the only people who would take that nonplan option are those who really want it, not those who just felt the choices were too complicated.

The other classic default option example is that of the organ donation checkbox. Many states give people the opportunity to identify as an organ donor on their driver's license. If the default is no, only a small portion of people will check the box to say yes. On the other hand, if the default is yes, only a small portion of people will check the box to say no. Presumably, this is because most people don't care. By changing the default option from no to yes, we can increase the number of people agreeing to be an organ donor without forcing anyone to do anything.

Not all nudges are positive. Marketing agents have known for years that creating Pavlovian drives can increase the likelihood that people will buy their product (sex sells). A classic case is the way the tobacco industry paid Hollywood to put cigarettes in movies to increase the number of smokers.[29] Advertisers can use that social celebrity relationship to sell products. This is the whole concept behind the influencer system that is now working on social media. People with a lot of social media followers can sell products merely by using those products in a video. Advertisers know this and are willing to pay a lot to these influencers.

Another classic example of a marketing nudge is providing a return policy. Because the Pavlovian (instinctual) system depends on immediate sensory stimuli, owning something, actually having it in your possession, increases the value you assign to it.[30] When it is still on the shelf or unclicked on the computer screen, one can distance oneself from the object, but once you have it, you are less likely to let it go. So companies say, "Try it for free, return it if you want." This is why car dealerships and house realtors always say, "Picture yourself in this car [house]." If they can get you thinking about it, they can activate your Pavlovian system and make you more likely to buy it.

Finally, we can identify this same process in the tax code. In the classic American system, employers withhold part of one's salary for taxes so that the employee never sees that taxed portion. In contrast, a self-employed worker receives the money and then has to pay it back. Anti-government, anti-tax groups have been pushing to remove the withholding taken from most salaried workers' paychecks because this will draw the workers' attention to it and reduce the willingness to pay that tax.[31] If the money never goes through your bank account, you never consider it yours, and you are more willing to pay your share into the pot. It will be interesting to see how the gig economy affects this dynamic. Uber, Lyft, MTurk, TaskRabbit, and other gig-economy contractors have shifted pay from paychecks with W-2 withholding to making individuals self-employed contractors who have to pay a separate self-employment tax.

The institutions we build will need to be designed to interact with our decision systems. Understanding how humans make decisions has consequences for our actions, which changes the moral decisions we make. We think that contingency management works by shifting people from a Pavlovian or procedural decision-making system into a more deliberative system (from an *Is it worth it?* decision into a *choose-between* decision). But it is not necessarily true that shifting to deliberative processes is always best. Everyone knows the danger of overthinking (being overdeliberative) when trying to be in the flow in a sports scenario such as performing a gymnastics routine or throwing a pass. In our discussions of the ultimatum game, we found that shifting a subject into a more deliberative decision-making process can make them more selfish, more "economically logical." In the Milgram experiment and the trolley problem, making someone more deliberative can make them more willing to sacrifice others for a "greater good." Furthermore, a lot of the decisions that we call moral depend on Pavlovian decision-making processes (like diving into the freezing river to save the drowning child). Sometimes the way to the assurance game is through letting our moral codes interact with our nondeliberative systems (such as how we can create empathy and reduce bias by teaching ourselves to take other's perspectives and see the world through their eyes[32]).

The important take-home message is that changing how we phrase the question has profound effects on policy and the consequences of that policy. One of the major things that moral codes do is to change the way we see a scenario, to change the question being asked. Changing the question can change which system is likely to be used in a given situation. It can change the actions that system takes in that given situation. We can build institutions (social norms, moral codes) to guide that behavior while maintaining individual freedom within it if we understand how those decision systems work.

14 JUSTICE AND REDEMPTION

The concept of justice is an important component of third-party social control. We will identify three kinds of justice, creating three separate moral structures: *retributive justice*, *restorative justice*, and *redemptive justice.*

In our discussion of social control (chapter 6), we talked about creating social control through several means, including first-, second-, and third-party social control.

Second-party punishment (if you cheat me, I will cheat you back; tit for tat) is the most straightforward, but it depends on extensive repeated interactions. If I'm only going to interact with you once, then there's no opportunity for me to punish you for cheating me.

One solution is to create aspirational goals and to convince the members of one's group that those aspirational goals are an important part of their internal narrative. That creates first-party social control as individuals attempt to ensure their own behavior is in line with those aspirational narratives. However, while first-party social control works surprisingly well, it doesn't work for everyone. Some people won't buy into those narratives and thus will play selfishly, perhaps even while espousing (verbally) those aspirational goals.

Another solution to this single-interaction problem that does not depend on first-party self-control is to form a community with reputation and third-party social control. However, reputation depends on learning about a person before interacting with them. This requires a tight-knit community where, even if I only interact with you as an individual once, I interact with people who interact with you. Third-party punishment requires a community that is tight enough to allow other people to punish you for your mistreatment of me. Third-party mediation depends on having a community that is tight enough for a third-party who knows both of us well to step in to negotiate a conflict situation.

As communities get larger, we need mechanisms to carry reputation beyond the small group, and we need laws to settle disputes between parties that come from separate communities that interact in only limited ways. Such tools include reputation markers, such as documents that can carry information across communities. For example, a state-issued driver's license is a reputation marker; it says that this person has passed the legal requirements to drive in one state. In the US, all of the other states are required to accept that reputational statement. As before, these laws can be used for good or ill. Michelle Alexander has made the strong case that even worse than the large (and racially biased) prison population in the US is the fact that most felons are legally marked for life and prevented from participating in many aspects of society (voting, taking jobs, getting bank loans), which prevents them from returning to society.[1]

This raises the issue of justice and the different goals of punishment and redemption. We will discuss three kinds of justice that interact with legal questions in important ways: *retributive justice*, in which an offender is punished for their crime, *restorative justice*, in which the offender is required to find a way to restore what was lost to the victim, and *redemptive justice*, in which the goal is to find a way for the offender to return to society.

Let's take a deeper look at each of these specifically.

RETRIBUTIVE JUSTICE

Retributive justice is punishment based on the concept of "you took from me, so we will take from you." Although there is strong evidence that humans have a Pavlovian desire for retributive justice,[2] theories suggest that the role it evolved to play as a moral code depends on two key hypotheses: First, that someone who is punished for a crime is less likely to commit it again and, second, that the knowledge that one will be punished for the crime makes one less likely to commit the crime in the first place. The first is an assumption about learning processes—punishment makes one less likely to take a given action again. The second is an assumption about economic processes—punishment increases the cost of an action and thus makes one less likely to take the action in the first place.

For example, if one is caught cheating on one's taxes, one has to pay all of the back tax plus a penalty. The penalty makes the cost of cheating higher than paying the taxes correctly in the first place. And while we hope that people pay taxes because they want to be a part of a cooperative group (i.e., first-party aspirational and social goals), we know there are individuals who won't. (Not paying taxes doesn't identify you as "smart"; it identifies you as selfish.) For those individuals, it is important to make the cost of cheating higher than simply not paying taxes because, otherwise, the selfish people won't pay taxes unless they are caught. Why would they? If they're caught, they pay the taxes. If not, they keep all the money they should have paid. So it's important to make the cost of cheating higher than the cost of paying in the first place. We can also note that tax penalties shift paying one's taxes from a prisoner's dilemma (better for the group if you pay your taxes, better for the individual if you don't because you might get away with it, and not getting away with it is no worse than paying the taxes in the first place) to an assurance game (better for the group if you pay your taxes but also better for you as an individual to pay just the original tax and not the penalty too).

When I was a graduate student and the teaching assistant for a computer coding class, we caught some of the students cheating. They had turned in code from another student, but the code still had comments in it stating "written by so-and-so" (who was not the student submitting the assignment!). I guess they assumed I would just run the code to see if it worked and never check to see if they had written it correctly? (This seems a strange assumption given that it was a how-to-write-good-code programming class.) The professor wanted to just give them a zero on the homework, but I argued that just giving them a zero made cheating equivalent to not turning anything in and that cheating was worse than not doing the assignment, so we should drop them a letter grade for the overall class.

At the University of Minnesota, if we catch someone cheating, we are now obligated to fill out a Scholastic Dishonesty Form and to identify the student to the Office for Community Standards. In part, this is because it was found that some students were very good at convincing multiple professors each time they were caught that it was a "first-time offense" and that they were very sorry and would never do it again. This rule that we have to follow is a good example of the importance of creating a legal process.

A legal process is a regulation that a community has found necessary to guide behavior. In this case, the older process (that the professor simply negotiates with the student) was found to be inadequate because as students moved from class to class they were able to cheat the system. Instead, a policy had to be created to catch these repeat offenders. (Importantly, I have found that the vast majority of students are honest and interested in their own education and recognize that cheating diminishes their own learning. I have encountered no more than one or two cheating students in twenty years of teaching at the University of Minnesota.)

On the other hand, many of these situations are not so small and deal with moral quandaries far beyond the grades of a few undergraduates in a computer science class. For example, the Catholic Church mishandled terribly the problem they had with pedophilic priests. In trying to protect the name of the Church itself, they made the decision to cover up the crimes and to shuffle accused priests from one parish to another or to temporarily remove them from a community, but they did not turn them over to secular authorities. (In part, this came from the idea that one could be redeemed by apologizing to God privately in the confessional rather than requiring a societal reckoning. We will address this issue when we discuss the moral structures provided by religion in chapter 15.) This decision turned out to be a major problem for the Catholic Church itself when the crimes and the cover-up were finally revealed. The cover-up not only allowed the crimes to continue (which was reprehensible in itself) but also made the Church hierarchy an accessory.[3] Notice how the moral structure implemented by the Catholic Church in response to the pedophilia problem was inadequate to address the problem and, eventually, blew up in their faces. A different choice of what rules to follow (what moral structure to impose) might have solved the problem sooner and reduced the suffering.

In short, the Catholic Church's moral structure here prioritized one part of the community over another—the priests over the victims. Similar problems of communities prioritizing an in-group membership over victims can be seen in the issue of sexual harassment in the workplace and the currently inadequate process for punishing and preventing the use of excessive force by police officers. All three of these situations have been marred by the ability to shift between legal groups without the records of one's past following one to the new group. Cheating within one group and then leaving it behind disrupts most of the available controls that we have to prevent that cheating; it disrupts our ability to enforce cooperation in the assurance game.

That being said, the idea that justice can follow from punishment alone is too simple. To really get to the two goals of retributive justice (that one should learn from that punishment and be less likely to take a given action again and that economic calculations should recognize that punishment increases the cost of an action and thus makes one less likely to take the action in the first place), we will need to take into account our decision-making systems (chapter 7). As we have noted, our different decision systems learn in different ways. This means that a learning-related justice goal should be aligned with those different decision systems.

Our instinctual/Pavlovian systems learn to associate stimuli and situations with emotional outcomes that release inherent action sequences. For example, an argument can trigger an angry response that could leave a person dead at the end of it. Processes would need to be put in place to dissociate that angry response from the situations that trigger it, and de-escalation processes would be needed to defuse it when it does arise. Simply putting such a person in jail would have little effect on reducing their future angry responses. (The violence of prison might well make it worse.) To use the analogy from chapter 8, we need to teach that person to feed the right dog.

Our procedural systems learn to associate arbitrary stimuli with a well-practiced action chain. A soldier having returned home can find themself responding to a nonlethal situation with dangerous (even lethal) responses. Processes would need to be put in place to extinguish the procedural responses and to associate those stimuli with different reactions. In fact, this is another example where understanding the neurophysiology is important. We know that extinguishing a procedural response does not entail actually unlearning or forgetting that earlier response. Instead, it depends on learning an override process that recognizes the change in the situation and context and enables a different response to come to the fore.[4] Moreover, procedural systems learn slowly, through lots of practice. A single experience is unlikely to change procedural learning.

Our deliberative systems work through a cognitive search-and-evaluate process in which we search through the consequences of our actions to determine the best option. Deliberation depends on an understanding of the structure of the world. For example, a person might work out that the likelihood they will go to jail for not paying taxes is low relative to the windfall they get for it or their ability to hire lawyers capable of talking their way out of it. To change deliberation systems, we need to convince the subject that the structure of the world is different and that their actions will have consequences.

While punishments can affect learning in all of these systems, the timing of those punishments, the number needed to change the learned responses, and the mechanisms of those processes all differ between the three systems. A long jail sentence is not going to have much of an effect on a Pavlovian response, and procedural learning requires many repeated examples, not one major punishment.

Current retributive legal systems are designed to confront deliberative decision errors more than Pavlovian or procedural ones. As such, many modern legal systems depend on the concept of "knowing the difference between right and wrong" and "having the intention to do wrong," which is known within the legal community as the concept of *mens rea*. The original goal is that we don't want to punish an accident. If I'm stopped at a stoplight and another car hits mine, pushing my car into yours and giving you whiplash, it's really not fair to punish me for hitting you. The specifics of mens rea and the importance of the constructs of "intention," "knowing," and "goals" and how they interact with the multiple decision-making systems is a task that (to my knowledge) has not been adequately addressed within the world of legal thought. However, it would be uncontroversial to say that retributive justice alone is inadequate as a solution to all of our community problems.

One of the issues with retributive justice are the Pavlovian motivations that drive a desire to punish those who have committed crimes. In part, this comes back to our evolutionary drives underlying tit-for-tat responses. This makes it politically difficult to find other solutions, but perhaps thinking of the goals of justice as a means of increasing cooperation within an assurance game can help guide some of those discussions. This leads us to two other ideas of justice that have been proposed as potential steps forward, that of restorative justice and that of redemptive justice.

The US faces a major problem right now in that, due in part to negative feedback loops, misjudged data from the 1990s (crime was already falling when the 1994 crime bill that increased punishments was passed[5]), and a private prison system making profits off of imprisoned citizens,[6] punishments have been escalating for minor crimes. Punishment that is disproportionate to the crime leads to the offender becoming the offended. The score is not "evened up." In a dyadic interaction, the offended party now has to defect in return, leading to a feud. In a community interaction, this can lead the offended party to respond by rejecting the community out of hand, possibly joining another community (committing treason, switching sides) or simply increasing their antisocial interactions if such an alternative community is not available.

Furthermore, these punishments are generally jail time, and there is strong evidence that prison, particularly American prisons, increase recidivism and violence rather than decrease it.[7] Importantly, many individuals also find their own way to redemption as well. A probabilistic description about a group should not be taken as a description about an individual. While 65 percent of people were willing to shock a stranger to extremes in Milgram's experiment, 35 percent were not (chapter 8). We would not want to punish everyone who participated in Milgram's experiment just for participating. The question we are addressing here is how to create policies that will shift that proportion, to decrease the 65 percent and increase the 35 percent. Moreover, subsequent postprison punishments for felons often limit opportunities for redemption. For example, it is hard in many states for felons to get jobs, car or home loans, or other opportunities. Some felons are required to announce their crime to the community before entering into a neighborhood, which certainly precludes any easy chance at redemption.

An important part of maintaining group cohesion is to increase the severity of punishment with repeated crimes. As we noted in our discussion of mesoeconomic games, sometimes an early warning that is not too punitive can return someone to the right track, but, as we saw in the day care example, one would not want to turn the punishment into a simple cost that the offender would be willing to pay. Studies of successful resource-maintaining societies (such as the ranchers studied by Ellickson, the lobstermen studied by Acheson, the sea-bass fishing communities studied by Ostrom, the studies of hunting and foraging tribes by Boehm, and the studies of Calvinist rules by Wilson) consistently find that early offenses are met with warnings, small punishments, and social disapproval, but repeated offenses are met with increasingly severe punishments, even unto ostracism or execution. This is an important part of community maintenance because we need to give the perpetrator of the crime a chance for redemption.[8]

The problem, of course, is that people overestimate the offenses against them and underestimate the offenses they commit. We have a Pavlovian desire to respond to threats with larger threats, to appear more aggrieved than we perhaps actually are, to overestimate the wrongs done to us, and to underestimate the wrongs we do to others.[9] Of course, this mismatch makes retributive justice difficult to align correctly and often leads to a vicious cycle of retribution and destruction.

RESTORATIVE JUSTICE

Given the problems of jail time, the difficulty of finding redemption from it, and the dangers of too much retribution, an interesting new process now being tried is the concept of restorative justice,[10] in which rather than throwing someone in jail as a punishment we force them to restore to the victim what they've lost or compensate society for the disruption caused. Sometimes even a symbolic restoration can be enough to de-escalate a retributive cycle.

In the movie *The Hundred-Foot Journey*, Indian immigrants settling into a small French town find their Indian restaurant in competition with a haute cuisine French restaurant across the street (the hundred feet). Some of the locals set fire to the Indian restaurant and write nasty graffiti on its outside wall. (The eldest son is able to put the fire out, but his hands are badly burned in the process.) In response, the owner of the French restaurant fires the sous-chef who participated in the attack and spends a day in the rain, scrubbing the graffiti off the wall. That act of restoration, albeit small, speaks as a visual apology—a statement that there can be a path back and a start to a journey that makes both of the restaurants better. (It's an assurance game!)

In chapter 6, we talked extensively about the process of "even up," whereby someone who overstepped provides a restoration in-kind, whether rebuilding a fence, providing a more valuable commodity in a trade deal, fixing damaged traps, loaning equipment, or even taking a side in an argument. In the repeating trustee games we saw in chapter 4, players coax trust back by providing more return on the investment.[11] (To remind the reader, in the trustee game one player gives money to another, the money increases in the transaction, and the second player can return some of the money back to the first player.) If the second player does not return enough money back to the first player, the first player will become less willing to provide money to the trust in future rounds. One way to repair that is to provide a particularly large return in a given round as a means of apologizing. Restorative justice is a way of forcing that repair process through legal means.

Reparations

A very interesting discussion is going on now in the US about whether the community of the United States (through the US government) should provide reparations to the descendants of slaves for their labor, pain, and suffering. The complexity of this lies in the fact that many current taxpayers did not derive their individual wealth from that slavery structure. Many of their families arrived as immigrants long after slavery was ended, and certainly those in the Union army sacrificed to end slavery. However, as pointed out by a number of authors, much of the US economic base was built on these enslaved workers, and many of the racial disparities that remain even to this day are due to racist rules and regulations from more modern times.[12]

The issue of reparations is complicated, in part because while many people in society may have benefited from inequality they did not themselves perpetrate crimes and in part because many survivors and descendants of victims may be uncomfortable putting a price on their relatives' lives.[13] Nevertheless, reparations can provide a means of apology and redemption, particularly when made at the level of societies. Making reparations at the level of the society can be seen in the light of third-party collective punishment.

As we talked about in chapter 6, there is a key advantage to making punishment come from the community rather than from an individual. Having the punishment come from the community helps sidestep the vicious cycle of retribution. Analogously, we can see reparations as a positive apology coming from the level of the society. In the punishment sphere, third-party punishment comes from people who are not the victims themselves. In criminal cases, police, judges, and lawyers are paid at the expense of the state, not the victims. In chapter 6, we noted a number of reasons to remove the victim from directing the punishment decision process itself. In the same way that third-party punishment of an individual says that the society condemns an action, reparations from a society (third-party compensation?) says that society apologizes for an action. Reparations provided from one group to another reduces both of these arguments against reparations. An individual paying taxes that lead to the reparations was not necessarily directly involved in the crime but agrees to participate in the reparations as part of solving the disruptions within a community. And reparations given to a community removes the one-to-one correspondence of the reparations as direct compensation for the loss of an individual life.

I've used literature for many of my examples throughout this book. There is a danger in doing so, however, because, of course, literature is made of stories written by writers interested in a literary goal, and these stories are not necessarily researched or cited, as a scientific work must be. Sometimes literary examples are factually wrong. A fascinating example of this is in the *West Wing* episode "Six Meetings before Lunch," in which the deputy chief of staff, Josh Lymon, who is Jewish, argues against paying reparations for slavery because his relatives were murdered in the Holocaust, and Germany isn't paying reparations. The thing about this is that Lymon is wrong here. (The writers of the episode are wrong here.) Germany has paid billions in reparations over the years to survivors of the Holocaust, to the World Jewish Congress, and to Israel itself. The German government continues to do so even to this day. In fact, the German government recently increased its commitment to these reparations, saying that the reparations are a way for current generations to make their disagreement with the past known, that current generations are united with survivors against the ideas of the past, and that the reparations are a way to prevent that past from ever recurring. It has allowed Germany and Israel to return to the positive non-zero-sum side of the assurance game.[14]

REDEMPTIVE JUSTICE

The perspective taken in this book is that justice is supposed to be a tool that societies have developed to increase the likelihood of achieving an assurance game in which people are more likely to cooperate with each other, achieving a non-zero-sum gain for both parties (for everyone in the community). As such, many have argued that the goal of justice is to return a person to the community.

We can come back to our example of the hunter in the tribe of central African foragers who hunted for large meat animals with nets. To remind the reader, this tribe hunted for large game animals with nets, and the perpetrator had put his nets secretly ahead of the others, thus stealing meat from them. They punished him by taking the excess meat and distributing it among the rest of the tribe (restorative justice) and also taking the hunter's family's store of meat (retributive justice). He apologized profusely, promised to never do it again, and admitted his error to the tribe. He was then allowed to continue helping in future hunts. They successfully returned him to the tribe's community to play the assurance game with them.

I got this story from Chris Boehm's *Moral Origins*. Importantly, however, Boehm also describes situations in which it was determined that the person was never going to participate in the assurance game in the future because of the severity of the crime, the crimes being repeated, or other evidence that the person was characteristically untrustworthy. In those other situations, Boehm describes more drastic actions, including expulsion from the community either as banishment or even execution.[15]

Guilt and Forgiveness

As I've defined it here, redemption is about returning the person who committed the offense (the crime, the sin) to the community. The goal is to coax them back to playing the assurance game with us. Several tools allow us to achieve that, including not only questions of restoration and punishment but also questions of guilt and forgiveness.

It is necessary to separate the legal term *guilt*, which means that one has committed a crime that the state identifies as unacceptable and will punish, from the sociological and psychological term *guilt*, which is an emotional consequence of a recognition that one has not lived up to one's aspirational goals. The religious term *guilt* is a combination of the two and reflects the aspirational goals embedded within most religions as supernatural interactions (*God will punish you; God will be disappointed in you*). Thus, the concept of religious guilt is an internalization of an external definition of a behavioral sin. (We will address these questions when we look at religion as a moral tool in chapter 15.)

Here, I want to address the question of guilt from a sociological and psychological perspective and note that the feeling of guilt is an example of first-party social control. Someone can feel guilty even if no one (but them) knows what they've done. In fact, psychological experiments suggest that emotional guilt is actually larger when someone thinks they've gotten away with something.[16] I suspect that this effect occurs because it opens the question of the community one is living in. *If this is a community where I can get away with this crime, then is this a community where others can too?* I think this manifests not in the internal psychological dialogue in this instrumental language but rather as a thought that one has not lived up to the kind of person one sees in oneself, that one has "let the family down," that one has "let oneself down," and that one has not lived up to one's aspirational goals.

Remember that there is a cost to third-party punishment. Third-party punishment is a form of tax on the community. (Often, it is quite literally a tax on the community—people pay taxes to fund third-party social control mechanisms.) Because of the non-zero-sum nature of the assurance game, if this tax makes people more likely to cooperate, then the return on investment may be positive. However, a well-built community is one where people are not committing those crimes in the first place. A society in which individuals feel guilt is one in which they are less likely to commit those crimes in the first place. So how does a society create these situations of first-party social control?

Many discussions of the question of guilt connect to the concept of forgiveness. Forgiveness itself can be a way of communicating that someone should feel guilt for an action they might not have originally seen as problematic. It may be a way of communicating to someone that they should have felt guilty and should ask for redemption. One of the most interesting issues that comes up in these discussions is that of nonviolence. Importantly, the statement that violence sometimes solves things does not mean that nonviolence doesn't also solve things. What is interesting here is to understand *why* nonviolence sometimes solves things. In large part, it is because nonviolence can show a strength of will and character that can drive aspirational first-party social control. In fact, one of the most fascinating things about guilt is the power of forgiveness to create guilt and lead to restoration.

The late John Lewis had a remarkable ability to get people to admit their guilt and beg his forgiveness for their crimes against him through his personal strength of character. Lewis (as he said) spent his life in "good trouble." He described how he knew that the hateful things done to him were based on an "illusion in the minds of those who hated [him]." It took many years, but his unwillingness to hate back made it more and more difficult for those who had attacked him to continue to hate him, created a feeling of guilt within them, and brought them to him to apologize and beg his forgiveness. As Lewis notes, however, forgiveness does not entail forgetting; instead it is about redemption.[17]

These questions of guilt and forgiveness occur with both devastating crimes (participating in apartheid, Nazi Germany, slavery, Jim Crow, redlining), and with small sins (stealing that last cookie from the cookie jar, lying to a spouse, a mean word said in the schoolyard). Fundamentally, however, the basic idea is the same. Guilt is a form of first-party social control created by having aspirational goals and finding that one's behavior did not live up to them. Forgiveness comes from recognition that redemption is possible; that it is possible to come back to the assurance game and work together in the future.

An important step in providing a path toward redemption is to understand the reason for the criminal act itself. (A wonderful *Saturday Morning Breakfast Cereal* comic defines heaven as "normal life, but everyone understands what your intentions were when you screwed up.") In chapter 10, we talked about the importance of empathy, of understanding the reasons why someone has taken certain actions. Sometimes that understanding can lead to a new path to returning the perpetrator to the assurance game.

One of my favorite TV shows is the series *Flashpoint*, about a Canadian police "strategic response unit" that tries to use negotiation and de-escalation tactics rather than force. In an early episode of the series, a father is incorrectly told that there is a heart available for his daughter who is on the list for a heart transplant. When they get to the hospital, however, they find that the daughter has been removed from the list, and they received the notification in error. So he grabs a gun and takes hostages, demanding they give the new heart to his daughter. The rookie on the police team sent to contain the situation casually refers to him as a "perp," but the senior member says, "He's not a perp. He's a father having a bad day." This change in perspective provides a means to de-escalate the situation, leaving the father, daughter, and hostages all alive and able to find ways to continue to contribute to society.

Importantly, it's one thing if someone makes a mistake, if one is a dad having a bad day trying to save his daughter, but it's another if the subject continues to commit crimes against the community. One way to address this is to escalate the punishment with continuing offenses. In studies of communities, whether they be Northern California cattle ranchers, fishers and farmers harvesting common-pool resources, or religious institutions, legal punishments grade upward with repeated defections.[18]

Increasing punishments only work, however, if they are aligned with the learning processes that underlie the decision-making systems that drove the crime in the first place. For example, a punishment applied months later is not going to be an effective learning signal for a Pavlovian crime of passion. As we've noted elsewhere, punishments also serve as a general deterrent, increasing the cost of committing the crime. General deterrents, however, only work if they are made public and applied evenly (i.e., not arbitrarily or unfairly).

Moreover, they only work for crimes in which there is sufficient planning, knowledge, and intention to take that increased economic cost into account. We talked in chapter 13, for example, about how increasing punishments and jail time are a particularly poor technology to decrease addictive behaviors because addicts are (by definition) inelastic to the costs of their addictions.

Interestingly, this comes back to the Samuel Butler *Erewhon* concept of treating crimes as diseases to be cured,[19] whereby we might consider Pavlovian and procedural transgressions as errors and dysfunctions in the decision process rather than crimes to be punished, per se. We have known for hundreds of years that physical changes to neural systems produce changes to personality and behavior. While much of the discussion of neuroscience and the law has been on whether to blame a subject for a crime if there is an identifiable brain abnormality,[20] the logic of that question is incompatible with our understanding of the science of decision-making. All decisions arise from the physical structures of our brains interacting with the external (social) world we live in. Some of those physical changes are more obvious than others, but increasing scientific technologies will enable us to recognize more and more of them. In my view, however, this question is irrelevant. What matters is what we can do with that knowledge.

We can see the usefulness of this medical understanding of neural systems in the very interesting case of a schoolteacher who suddenly developed highly sexualized behaviors, collecting child pornography, propositioning the girls in his school, and taking on other unacceptable and inappropriate behaviors. He was found to have a glioblastoma (a tumor in the brain). When it was removed, his behavior abated. When it regrew back a year later, his behavior returned, and further surgery removed the behavior again. One option, for example, would be that instead of throwing this person in jail one could allow him probation and require him to get regular scans to determine if the tumor is returning. As long as he got checked for the tumor regularly, he could safely be allowed to remain in society. However, if he failed to attend those regular checkups, he would be liable for his actions and punished retributively.[21] This is an interesting example of increased cost due to the threat of retributive punishment providing incentive to do the work of self-regulation.

I have defined redemption as a return to the community; however, some may suggest that redemption requires an admission of guilt, an internal sense of remorse, and requests for forgiveness. Furthermore, most concepts of redemption require some statement of accountability. Forgiveness does not imply that one forgets the act, only that one allows the actor back into the community.

Many programs that attempt to modify behaviors (for example, in addiction) talk of patients being in "recovery." This word was, of course, chosen because of its medicalized perspective instead of the moralistic perspective of "redemption" because these programs want to see addiction as a disease to be treated rather than as a moral failure to be punished. The new more medicalized views on crime and justice may suggest that we should start thinking of these questions in terms of the process of recovery, whether that be of an individual recovering a place in society or society recovering from a tumultuous situation. One key, of course, is that these treatment programs see recovery as a continuous process that requires monitoring and continued attention.

As we have seen in the discussion above, all of these constructs play a role in the ability of a subject to return to a community. I would argue that, just as the concept of redemption is a tool to achieve the assurance game, these constructs are tools to achieve redemption. So, yes, redemption often requires an admission of guilt, an internal sense of remorse, and requests for forgiveness, but primarily because those moral tools are what enable a return to the community.

15 RELIGION

Religion provides a social technology that defines groups, offers a sense of being a part of something larger than oneself, enforces third-party punishment, and avails us ways to ensure that the members of our community cannot escape observation of their sins.

When I was asking colleagues to read early drafts of this book, one of them asked: "If a new moral structure improves the lives of more people within a community but makes them believe in God less, is that a success or a failure of morality?"

For millennia, religion has been seen as a key factor in morality. In his book *Rocks of Ages*, Stephen Jay Gould argued that science and religion encompass "non-overlapping magisteria," with science being about facts and religion being about values. This was an attempt to placate the anti-science sentiments argued by many in the religious communities, particularly in the US. The idea was that science and religion could coexist because they work in separate fields. Gould argued that science was about description—*how the world works*, while religion was about prescription—*what one should do*.

However, contrary to Gould's original claim, both science and religion make claims about both description and prescription. In chapter 12, we not only saw that descriptive knowledge has prescriptive consequences but also that science itself has prescriptive goals. Similarly, religion is not solely about prescription; it contains description within it as well. A surprising number of religious statements are descriptive: *The earth is five thousand years old. Humans and other animals are not genetically related. If you misbehave, a god will punish you. Your dead ancestors are watching you, so don't embarrass them.* Furthermore, for many religions some prescriptions seem arbitrary. On the other hand, many of the prescriptions across religions are also remarkably similar. These similarities suggest that while religion may be serving a purpose of helping us achieve a prescription, it is a prescription that lies outside the realm of the religion itself.

Given the definitions that the goal of morality is to drive us into the assurance game and ensure that we are cooperating with a group of cooperators, then perhaps religion is a tool helping us do that. If so, then neither religion itself nor its proclamations would be the goal of morality. Notice the very important difference between my colleague's question and the statement by the pastor I met on the plane who said he wasn't worried because any new science of morality would find that religion had been right all along about what makes some behaviors moral. The pastor was suggesting that religion was accessing a deeper fundamental question of morality even if it wasn't the source of morality in the first place.

Notice also that there is a big difference between my colleague's question and the suggestion "You will find religion is necessary to get people to behave morally." My colleague's question is about the prescriptive goals of morality—*Is belief in God the goal, or is the goal something else?* In this book, I have taken the stand that the goal is elsewhere (creating assurance games). In contrast to my colleague's question (which is about the prescription of morality), the suggestion that religion is necessary to behave morally is actually a descriptive scientific question about behavior. It is a hypothesis about the social structures that work to create assurance games. And while the hypothesis that religion is necessary to make people moral is a viable theory within the science of morality, the evidence against it is pretty strong—people without religion still behave morally, and cultures have been able to create aspirational goals and first-party social control without religion.[1]

I am not going to take on the large question of *What is religion?*, neither from a religious point of view nor from a secular one. Furthermore, throughout the history of religion there have been arguments about how much one can separate out the various descriptive and prescriptive religious components from each other, such as spirituality, faith, or the existence of supernatural beings. These questions are beyond the scope of this book and would take us into a discussion space that I am certainly not qualified to address. Similarly, I am not going to address the question of whether religious statements about the world are factual or not. Many other books address those issues, and the discussion would take us away from our goal. What I want to examine here is the point that the pastor on the airplane made: that some of the things religion provides us are aspects of a social structure to help create morality. So, for now let us take religion as an institution created by humans and ask *How does religion affect moral questions? How does it change the assurance game?*

CREATING GROUPS

The most obvious explanation of religion is that it defines a group. Several researchers, most notably David Sloan Wilson in his book *Darwin's Cathedral*, have argued that religion is a means of defining a group. At the simplest level, there are others who go to the same church, synagogue, temple, mosque, or house of worship as you. These are friends; they provide support. In many societies, these communities provide resources to people in difficult times, the congregation works together to support each other, and the other members serve as a safety net and insurance policy. But Wilson argues that religion provides more than this. Religion provides a society beyond the immediate group of known friends.

Many religions require complex procedures to join, including both emotional statements of renunciation of the self ("I accept this role that God has provided for me") and physical trials (such as symbolic drowning or physical mutilation). Importantly, these procedures usually require renunciation of all other religions ("There is no God but Allah"), meaning that (for many religions) one cannot be part of multiple religious groups at one time. These complex procedures serve two roles. They make it difficult to join, which limits casual participation. But also, humans are more willing to go the extra mile for a group they have worked hard to become a part of.[2]

Moreover, these procedures are generally public. According to this hypothesis, an important part of joining a religion is the capability of recognizing others within the religion. Thus, the public definition—one needs a way to recognize who is a part of the group. Even when groups are secretive, they have symbols that allow them to recognize each other—code words, secret handshakes, seemingly innocuous behaviors that only those within the group understand. Religion defines a group, with the understanding that one will cooperate with others in that group.

A SENSE OF PURPOSE

Many religious individuals say that their religion provides them with comfort and an internal sense of purpose. We can, again, take that as a scientific observation. Why is it that faith provides this sense of comfort to many?

When it comes to human teamwork, we don't want to be just a part of the team; we want to play a role within that team. Emotionally, this is identifiable as the need to find a purpose. For many people, religion provides this. But, as with the other things religion provides, religion is not special here—other means and other mechanisms can provide that higher purpose. We see this in social justice movements, in national revolutionary movements, and in science itself.

In writing this book, I've realized how much of my own life has been built on "contributing to the scientific edifice." In my view, science is a grand journey that humanity is taking. My role is to put a brick in that wall. I'd never thought of this as a faith before, but I see now how that view of my role in a journey larger than myself has provided a lot of my own personal resilience and stability.

One of the tenets of Judaism, particularly modern Judaism, is that of *tikkun olam*, which translates roughly as "repair of the world." The concept is that the world is broken, and humanity's job is to fix it. But, most importantly, *tikkun olam* recognizes that repairing the world is a difficult problem that cannot be solved by one person alone. It is a multigenerational project. And so the other key is that "it is not your role to finish the task, but you are not free to desist from it either."[3] Basically, *tikkun olam* is saying that you need to make progress. If you make progress, the team will see it through. *Tikkun olam* is about finding one's role in a larger story.

In our early examples of the *Golden Balls* games (chapter 2), we encountered Michelle, who was playing split-or-steal with Ella. Michelle decided to cooperate (split) from the beginning, citing religious reasons for her decision. When she suspected that Ella had chosen the steal ball, she hesitated, but again came back to "The Bible says thou shalt not steal, so I'm not going back on that." What I find fascinating about this interaction is that when Michelle discovered that Ella picked steal, she was angry, yes, but she was also clearly disappointed in Ella. In Michelle's view she had given Ella the opportunity to be saved and Ella had failed. One shouldn't steal and Ella had violated that rule. In Michelle's view, sharing was the right choice and even afterward, she said, "Regardless, I wouldn't have stolen anything," and "I'm still blessed."

One of the major things that religion provides is a code of conduct, an aspirational target that tends to be supportive of a community and supportive of cooperating. These codes tend to be immutable absolutes that do not accommodate nuance. Several researchers have pointed out that immutable absolutes are like setting the value of choosing that option to infinity, which prevents negotiation or pricing that goal away. Identifying something as absolute and sacred makes it easier to make deliberation choices—one doesn't need to ask how much one is willing to take to give up that goal because that goal is beyond price. In fact, people often become offended at the idea of being offered a price to weaken something they consider sacred.[4] By practicing these goals regularly, one can even make them an automated (procedural) habit. Procedural learning depends on the regularity and stability of the decision. Making decisions based on an immutable absolute makes them stable over time and conditions and thus encourages the creation of a proceduralized habit of following that moral structure. (Of course, habits are also hard to break as moral structures change, but under stable conditions, procedural habits can be advantageous.)

Immutable absolutes are a key factor in aspirational goals and help provide for first-party self-control. These aspirational goals provide support for one to avoid temptations by defining one's self as "the kind of person who does not do that" and also by providing a disappointed third party (whether it be a god or one's ancestors or merely the other members of one's community). These goals access Haidt's purity construct of morality, as someone who has failed to maintain that self-control has "sinned," is no longer "pure," and is no longer "blessed."

Of course, actually achieving perfect self-control is extremely difficult (in large part because willpower takes effort and is hard to sustain, leading one unto temptation[5]), so most religions also have redemption processes that can return one to the fold to repurify the sinner through appropriate rituals and confessions.

THE QUESTION OF PURITY

Purity in general is a major tool of religions. The concept of purity is a definition of one's success in meeting the set of aspirational goals within the religion, thus one who meets a set of criteria, which often includes following a set of rules, is "pure," and one who does not is "impure." Purity creates a set of aspirational goals to achieve.

In many religions, however, purity goes beyond following a set of sharing and cooperation rules. For one thing, it often includes a set of complicated and seemingly arbitrary rules, such as a ban on eating certain foods or an insistence on performing certain rituals at given times of the day or after specific events. It can even include definitions that are immutable, such as ethnicity, caste, or skin color. As such, the construct of purity can also be extremely dangerous, as it can exacerbate the problem of parochial altruism.

Purity is a way of manipulating the intrinsic goals of humans by attaching negative (disgust) feelings to things and behaviors in the world. Humans are intrinsically disgusted by unhealthy conditions—feces, rotting food, blood— and generally react to these things with revulsion. The moral construct of purity, which Haidt points out is the opposite of disgust, is a tool to identify the existence of additional unacceptable things in the world that should disgust us.[6]

As such, we can understand the tool of purity in terms of Pavlovian decision processes. There are disgust-related behaviors (revulsion, aversion) that are intrinsic (Pavlovian) responses in humans (see chapter 7). Pavlovian learning associates stimuli and situations with intrinsic responses. The classic example in psychology is the creation of a fear response to an arbitrary stimulus (a tone or an object or a place) by pairing that stimulus with a painful shock or the creation of desire by pairing that stimulus with a positive reward (food, sex, positive social interactions such as grooming).[7] In a similar way, the concept of purity associates some arbitrary actions and things with good rewards (heaven, music, social appreciation) and other arbitrary actions and things with bad punishments (hell, pain, ostracism, feces, and rot).

Two important issues make purity both a powerful and dangerous tool. First, purity tends to provide sharp distinctions. Most religious definitions of purity make sharp distinctions between pure and not, in the group and out, good and bad. Second, many religions use arbitrary definitions for purity, including questions of family lines, skin color, or other factors that can be hard to modify, making those lines hard to cross. Once you've declared another person impure, it can be hard to recognize that other person as even human.[8] Sharp distinctions make it easier to stay within the lines, but they can also increase the difference between the inside and the outside of those lines. Sharp distinctions in group definitions can increase the xenophobic side of parochial altruism and make one less willing to compromise with the impure.

As with other religious tools, we can see the idea of purity enacted in other facets of our social toolbox, such as nationalities, families, and other communities. Again, the dangers of purity come with the rigidity of its definitions. The impure are defined as *other*, frightening, dangerous, and competitive. They are assumed to be selfish and likely to defect against you, which, given the rules of the assurance game, means you should defect against them. And, just as purity can be defined by factors outside of one's control in religious perspectives (such as family lines), purity in nonreligious facets can be defined by skin color and caste, making it a dangerous tool for group definitions.

On the other hand, religion and a sharp insistence on maintaining the purity of one's own certain convictions can also be a tool for strength in the face of devastation. Aimed correctly, that strength can lead to forgiveness and a means of transforming outsiders into the fold. In *Across That Bridge*, the late John Lewis spoke about how his faith provided him with the strength to take actions he knew were right because he knew in his heart that good would eventually prevail. He talked eloquently about how maintaining that purity of nonviolence enabled him to reach across and bring others to his side. His description of how the man who beat him as he got off a Freedom Rider bus in Rock Hill, South Carolina, in 1961 came to ask his forgiveness years later is a testament to the importance of strength, faith, and the maintenance of aspirational goals. As we discussed in chapter 14, there is a complex interplay between forgiveness, accountability, the acknowledgment of, and the remembering (or forgetting) of a sin.

THIRD-PARTY PUNISHMENT: SIN, SHAME, AND REDEMPTION

As we saw in chapter 3, the tax game contains the fundamental problem that the selfish within that group taking advantage of the generosity of the group outcompete the cooperators within that group. Most religions include a central concept of "sin"—an inability to live up to the rules set out by the religion. We can divide sins into two categories: sins against the community and sins that violate seemingly arbitrary rules. Let's work these two categories backward.

Violating Arbitrary Rules

Every religion seems to have a set of strange rules that do not translate across religions. Many of these rules do not have obvious sources, and while just-so stories can be given as reasons, those just-so stories almost never hold up under scrutiny. For example, classical Judaism has a strict set of food preparation rules ("keeping kosher") that include not eating pork, keeping milk and meat separate, and allowing only certain kinds of cicadas to be eaten. While some researchers have suggested that early versions of these restrictions may have arisen from health-related observations, further studies have generally found that the rules are more based in tradition than in logic and suggest a process of defining purity rather than of protecting health.[9]

Moral judgments of these arbitrary rules tend to be made in a binary manner and are intolerant of mitigation.[10] This is to be expected if the measure of these arbitrary rules is a moral judgment of the definition of the group—people who conform to the arbitrary rules are a part of the group, and people who are unwilling to conform are not.

That definition is almost certainly the key to the purpose of these arbitrary rules. They define a community. Are you willing to put up with these arbitrary rules to mark yourself as part of the community? If you are, then I can trust that you will remain part of that community with me. If not, then maybe you shouldn't be part of my community at all. Consistent with that logic, the rejection of these arbitrary rules tends to be punished with ostracism—that is, rejection from the community.[11]

While each religion does seem to have a set of arbitrary rules that define its community (don't eat pork, do eat fish on Friday, don't use electricity, don't cut your hair), most religions also have sets of rules that are common across religions. Perhaps these more common rules are less arbitrary. Many antireligious sentiments are based on the observation that since some rules are clearly arbitrary and seem to serve no purpose (as noted above, one potential purpose may be to identify groups), then all of the rules must be arbitrary. However, if we look more carefully, we can find moral explanations for the rules that are common across religions.

Sins against the Community

The key to successful communities is within-community cooperation. Almost all religions contain a set of rules that help maintain within-community cooperation. Many authors have noted that the underlying tenet of almost all religions is the Golden Rule: "Do unto others as you would have them do unto you" (sometimes stated as "Don't do unto others what you would not have them do unto you"), which appears across religions in both the East and the West.

There is a classic joke that the real Golden Rule is "He who has the gold makes the rules," which, I would point out, is also a moral statement. What is interesting is that in the evolutionary competition of memes and institutions, most humans generally reject communities based on the transactional, hierarchical Golden Rule (Rule of Gold?) in favor of communities that at least give lip service to the more traditional Golden Rule. This is, of course, because the more traditional Golden Rule ("Do unto others as you would have them do unto you") is a cooperator's community, while the joke is not. This is the point of the science of morality. Some of these moral statements increase our ability to work as cooperators within a group. Others do not.

Most religions, however, also provide a set of behavior rules beyond the Golden Rule that can include remarkable similarities. There is an old story that Hillel (a Jewish philosopher in Roman times, 110 BCE–10 BCE) was asked by a Roman soldier to describe his entire religion while standing on one foot. Hillel replied, "That which is hateful to you, do not do to others. All the rest is just commentary. Go and learn." The first part is of course the Golden Rule. What is interesting is that the commentary is also surprisingly similar between religions.

Some of the obvious ones include "Thou shalt not kill" (which is a good basic rule within the community that should reduce intracommunity conflict), but others include definitions of sexual relationships, fiscal relationships, the protection of children and the weaker members of the community, and within-community roles.

Obviously, a keen source of intracommunity conflict is sexual interactions. Raising a human child requires a tremendous amount of investment, both personal (in the sense that it takes a lot of time out of one's day and a lot of emotional effort to train a child to be a successful member of a community) and financial (in ancient worlds, this meant hunting or gathering more food and other resources; in the modern world, this is often literal financial capital). Knowing who a child's parents were was often an important factor. As noted in chapter 9, we have evolved intrinsic goals that are correlated with genetic success. Given the lack of DNA testing available to the evolutionary process, this desire to have both parents provide reliable investments in their children drove the evolution of sexual jealousies that can create dangerous situations of sexual conflict and tear communities apart.

To reduce sexual conflict, most religions provide some definition of a sanctioned (marriage) relationship in which two people define themselves as a specific, separate, and identifiable pair. Marriage is actually a contract between three parties: the couple and the community that agrees to treat them differently. This means that, importantly, marriage provides rights and responsibilities between partners beyond reproduction and the raising of children. In the US, these include things like joint financial accounting, the right to be considered a single entity for tax contributions, the right to make end-of-life or hospitalization decisions, and other social definitions.

Typically in the modern world, we think of the sanctioned relationship as a two-partner construction of limited sexual interactions providing the mechanism to maintain investment in one's genetic offspring, but other community constructions are possible. In the current modern society, many children are raised by single parents (hopefully with some community support from grandparents or friends, as raising a child truly alone is an exhausting and difficult task), and many parental pairs live apart or remain unmarried while still raising children together, in part because they are unwilling to put themselves into the joint structures defined by marriage, such as merging finances, taxes, and other relationship components. Many ancient cultures included polygamy, polyandry, and other less dyadic relationships. In many cultures, the extended family plays a particularly large role in raising children. In human society, marriage and religion are often entangled, but that is not because marriage is a religious act; it is because religion has been the way we've organized communities for a long time. Marriage is a community act.

Other sources of intracommunity conflict include fiscal relationships, and many religions provide detailed rules for the reduction of fiscally driven intracommunity conflict. The classic example being, of course, "Thou shalt not steal." But many religions go far beyond that. For example, many religions limit usury and predatory lending practices. (Islam, for example, rejects the concept of interest, which has led to fascinating processes to sidestep these rules to allow modern finance and banking systems to thrive within the context of such laws.[12]) Many religions include *jubilee years* in which debts must be forgiven.[13] And many religions include an expectation of helping neighbors in need, a central insurance structure based on tithing to the religious community, and rules for providing resources to the weak and poor.

How to handle questions of children in particular and the weak and the poor in general is an important part of any community. The key, of course, is that children grow up to become members of the community, poverty can be a temporary situation, and people who have limited abilities can still play many roles, freeing up more complex roles for those with specialized abilities. Moreover, because of the way human Pavlovian systems learn, teaching one to treat all humans well, including the weakest among us, can percolate into the rest of the social sphere, improving the success of cooperative communities.

In his discussion of religion and community (*Darwin's Cathedral*), Wilson talks about how early Christian groups survived the Antonine and Cyprian plagues better because they risked helping people within the community who were ill. Treatment can make an illness temporary (thus returning the ill person back to the community), while lack of treatment can make the effect of illness permanent (due to death, which doesn't return the person back to the community). Similarly, Wilson talks about Korean Church communities in Texas that helped immigrants arriving with limited resources start businesses and become full-fledged members of the community, which could then be "paid forward" to new immigrants. Obviously, this intracommunity financial, physical, and emotional support provides for more successful communities than leaving everyone to fend for themselves on their own.

I don't want to be a Pollyanna about religion—these religious roles can be stifling and limiting. Many religions have a hierarchy that places certain members (priests, castes, families) above others and identify sins against that hierarchy as particularly egregious. Hindu caste systems limit opportunities for those within lower-ranked castes to achieve higher-ranked positions by promising better opportunities in the next life (for which there is no actual evidence). Many religions define sex-based roles that are tremendously limiting. Religions often punish sexual infidelity by men and women very differently, and rules such as requiring women to wear a burka put the onus of sexual purity on one subgroup over another.

Of course, our between-group competition implies that religions will compete for the system that provides the best opportunities that people want the most. However, as I have noted elsewhere, these processes are optimizing within the current sociological ecosystem; we cannot assume that this cross-group competition is complete. Thus, we should not assume that the various surviving religions are "optimal" solutions to this cross-group competition. We will come back to cross-institution comparisons in chapters 18 and 19 when we address issues of moral relativism and institutional design. It is very possible that religion has outlived its usefulness, and there are newer, more modern structures that can provide moral solutions with less baggage.

Punishment

If the aspirational goals fail, then we may still find people committing selfish acts. It is important to the community that those selfish acts that do occur do not empower the perpetrator. The key to preventing a selfish act from empowering someone at the expense of the community is to reduce the success of that selfish act. The obvious solution is to punish that selfish act, to change the rules of the game to make it less valuable (chapter 6). Religions often provide a process of third-party punishment. If the goal is to return the person back to the community, the key is to construct third-party punishment processes that escalate with continued sins. Most religions provide explicit punishments for sins. Sometimes those punishments exist within the community (such as ostracism from the community or a public apology), but more often they are punishments that arise beyond the present. For example, the classic idea of heaven and hell is punishment for sins, as is the statement that what you do in this life changes what role you play in the next.

We've already seen how gossip can provide a means to bypass the problem of dyadic interactions and allow the community as a whole to punish selfish acts, but even gossip depends on someone starting the chain. If no one observes the sin, then there is no way for the community to punish that crime. In order to punish a sin, one has to catch the sinner and recognize that a crime has been committed. One solution is to convince the members of your group that someone is always watching—that there is no way to avoid detection of your sin.

Most religions structure a belief in the world that someone is always watching, whether "God is everywhere" (Christianity), "The Gods see all" (Hinduism), or that one's ancestors are watching over one, as, for example, suggested by Confucianism. Even religions that do not include an explicit omnisciently observing agent suggest that one's future is judged by the complete sum of one's life's actions, including those not seen (e.g., weighing one's heart against a feather in the Egyptian *Book of the Dead* or the Hindu and Buddhist concepts of reincarnation, in which the sins committed in this life determine what one is reborn as in the next life). Peter Boyer has suggested that this overwatching is one of the original underlying purposes of religions that has made them successful. Individuals who believe in a religion with an overwatching process would be more likely to play fairly at the group level (to be more cooperative, using the definitions we defined earlier) and less likely to cheat.

Redemption

A key factor in any religion is the possibility of redemption. While some individuals may be unredeemable and require a permanent solution to remove them from the community (by ostracism or death), a community that enables an individual to make restitution and redeem themself will outperform a community that punishes every sin with a permanent solution. If a community can turn a selfish sinner into a productive cooperator, then that community will have one more cooperator participating in the non-zero-sum assurance game.

Redemption is a process of return. If we look at religions across the world, we find that redemption generally entails a public apology and a recognition of one's sin, some form of restitution, and a promise not to sin again. These processes make sense from a community construction perspective. The public apology and recognition of one's sin (particularly with the concept of remorse) suggest that the promise not to sin again is sincere and that the sinner is unlikely to repeat their mistake. The restitution helps undo the consequences of selfishism and is a form of restorative justice. Remember, the key problem we are trying to solve is that the selfish outperform the cooperators within a group, and our goal is to make it advantageous to make cooperative rather than selfish choices in general. The public nature of the admission and apology provides warnings to other potential sinners and prevents sinners from switching groups to hide their sins.

Interestingly, the concept that God is always watching and that punishment will be meted out in the afterlife led Catholicism to the idea that a private apology to God is sufficient punishment and that one need not make a public apology for one's sins. This turned out to be a major mistake from a community-construction perspective, as can be seen recently in the Catholic pedophilia scandal, in which pedophiles (who commit sins against children and should, most cultures agree, be severely punished) were able to apologize privately and were shunted to other dioceses where many relapsed back into their sins. The Catholic Church is still attempting to deal with the fallout from its mistaken punishment plan.[14]

The problem of transferring between groups to hide one's sins has been a problem in several institutions, including sexual predators in academia, violent and racist cops in police forces, and pedophilic priests. These problems interact with issues of subgroups and power structures and how those subgroups protect their members. (This is related to the issues brought up in the very first chapters that a system needs to have moral and social codes to deal with "the enemy within"—a subgroup that protects itself at the expense of the larger community.)

Recent changes in communication technologies have made it harder for these individuals to escape their past sins. These changes have also allowed communities to recognize victims, leading to alterations in the moral views of how to handle sins by members of powerful subgroup communities. These revelations of pervasive problems within powerful subgroups can lead to a lack of trust among the larger communities, with devastating consequences on the social capital (social structure) within that community. The journey out of the problematic structure (in which a powerful subgroup protects sinning perpetrators) into a society based more on the assurance game can be a long and difficult one.

Importantly, punishment, whether it be religious or otherwise, usually entails escalating consequences with repeated sins. Escalating consequences not only provide multiple opportunities to return to the community but also prevent the economic trade that can arise when each sin has a simple price. It is important when constructing third-party punishment to create an appropriate escalation process through which early sins (which are often mistakes) can be recovered from but repeated sins (which are often intentional selfish acts) do not avail the sinner the opportunity to take advantage of the community by simply paying the initial punishments. Moreover, third-party punishments need to be seen as community-driven and not simply transactional costs to decrease the value of selfish acts.

Most religions include a process of redemption that depends on confession, restitution, and remorse, allowing sinners to return to the fold. Escalating punishments prevent selfish actors from taking advantage of a community. Incomplete punishments, however, can disrupt and destroy communities from the inside.

One of the most fascinating things I've found in working through these ideas, and in the writing of this book, is that, quite literally, religion really is about morality. It defines groups through shared beliefs, shared rituals, and shared sacrifice. It defines structure within those rules so that everyone has a part to play. It embodies rules that limit intragroup conflict.

Let's take, as a canonical example, the Ten Commandments in the Abrahamic religions (Judaism, Christianity, Islam). In the King James Bible, which has defined them for a very large group of humans, the first three are about defining the group—for example, the first is "Thou shalt have no other gods before me." Note the logic here that implies that there are other groups that have other gods. This group is defined by the one god they will worship. The second and third ("No graven images"; "No bowing down to others") separate this group from the other groups. Since other competing religions, particularly when these were originally laid out four millennia ago, often defined themselves through idols and images and rituals to those images, these two commandments meant that a group member couldn't be part of those other groups. The fourth ("Rest on the Sabbath day") provides both group definition—the group members are the people who rest on a specific day of the week—and also enhances in-group cohesion, preventing one from working anyone else within the group 24/7. The fifth ("Honor thy father and mother") defines structure within the group, which was extremely nuclear-family oriented. Note that I am not defending any religion or this structure. I am merely observing how one of the most common examples of religious doctrine is really all about defining and maintaining group structure. Completing the set, the sixth through tenth are about reducing intragroup conflict ("Don't kill"; "Don't sleep with someone else's spouse"; "Don't steal"; "Don't bear false witness"; "Don't covet").

The last two are interesting because they are about thought, not actions. While the ninth commandment is often interpreted as "Don't tell lies," many scholars (anthropologists, sociologists) have suggested that the "Don't bear false witness" text should be taken more literally, as "Don't tell wrong gossip."[15] We've seen throughout this book that gossip is an important community control, shifting dyadic interactions with limited information into group interactions and shifting single-interaction prisoner's dilemma games into repeated-interaction assurance games. However, as every teenager knows, it is possible to destroy someone's reputation through false gossip. Some have suggested that the ninth commandment is really about preventing reputation destruction by "bearing false witness."

The final commandment is about accepting each other's private property (which in the original formulation included the male citizen's wife and servants but whom we would certainly not include in a modern formulation of society: much of the historical transformation over time has been the inclusion of more and more individuals into the realm of humanity with an independent agency). Getting your flock to not covet each other's possessions would be a very good way to reduce the tension that comes from income inequality. Whether we in the modern world are willing to accept this formulation or not, there is no question that the tenth commandment was designed as a means of reducing intracommunity conflict.

It is intriguing that if we look across religions, we find that all religions embody similar sets of rules—defining what it takes to be a member of the group, defining what the roles are within the group, and providing regulations that reduce intragroup conflict. In *Darwin's Cathedral*, Wilson looks at the rules written down by John Calvin in 1538 and comes to a similar conclusion, identifying rules that provide for explicit within-group role structure (parents over children, magistrates and pastors over the population), explicit exhortations to support the community over self ("Abandon self-will," "Behave as a single organism," "Pay taxes"), and structures to reduce intragroup conflict ("Don't harm or kill your neighbor," "No theft, either by violence or cunning," "Don't lie," "Behave in a civil manner").

In general, all studies of religious rules find similar exhortations. However, these studies also find a host of other, arbitrary, rules that do not necessarily address group-control questions. For example, is the religious belief that the earth is only five thousand years old a critical part of that social structure construction? Should that belief be seen as a group-construction tool (as a marker like a tattoo), as a mental object that has infected the cognition of a population, or as an epiphenomenon of the complex of concepts that make up the religion? What happens when the data disprove that belief? Does the whole religion collapse, or are they separable components? These questions are beyond the scope of this book. That being said, I think the data are pretty clear that religion has played an important (at least historical) role in the way that we have created and controlled ourselves in groups. But that doesn't mean that it is the only way.

16 GOVERNMENT

Government, like religion, defines groups, provides a sense of being a part of something larger than oneself, provides for aspirational goals, enforces third-party punishment, and avails us ways to ensure that the members of our community cannot escape observation of their sins.

The last chapter suggested that religion provides a form of government. Government has long been described as a means of coordinating groups to provide non-zero-sum gains and as a means of controlling groups to provide for the power of one group over another. In the language used here, government is a way of formalizing the rules of the road and social norms.

We have seen that religion provides a means to help groups form, defining the group itself and enforcing third-party punishment, as well as a means of self-government, allowing the individual to govern themself to avoid the selfish behavior that breaks groups apart. One reason that individuals within a religion take moral actions is because it is part of their self-identity and because the religion provides them with an aspirational moral structure to help guide decision-making (religions provide punishment and incentive changes to guide behavior as well). Interestingly, we can see all of these properties in formal governments, including the same aspirational moral structures that define self-identity and that help us take moral actions driven by internal controls.

An interesting question is where to draw the line between government and social norms. For example, small communities tend to have well-established but unwritten social norms that they use to govern themselves. Sometimes these social norms are not even explicitly identifiable by the community members when they are asked about them.[1] But as communities get larger, it becomes more and more difficult to maintain that internal structure as the interactions become sparser within that community. In this chapter, I am going to examine governments as the formalization that provides a means of dealing with those sparser interactions.

However, in the end, governments are consensus fictions pretended into place;[2] they depend on the consent of the governed. Sometimes that consent is held at the end of a sword or a gun, but such consent can be fragile as the oppressed rise up in revolution or join an opposing group that is more willing to provide opportunities. Since oppressive consent is fragile, governments, like religions, tend instead to try to achieve that loyalty through first-person aspirational goals.

CREATING GROUPS

Modern governments identify group members through documents (such as one's passport, driver's license, or other identification) and separate individuals into specific memberships with certain rights and responsibilities. While some nations allow dual citizenship, most do not, and those that do are generally close allies, working as part of a larger community. Even within a federated nation like the US, one is not supposed to hold driver's licenses from multiple states. As communities become more complex, more specific potential roles within the community become identified. Modern governments identify multiple roles within their population—the US differentiates citizens (with the right to vote and the responsibility to sign up for a potential draft), green card holders (who have the right to work and live within the US), and various kinds of visas that allow specific roles to be played within the population (work visas, student visas, tourist visas). Similar differences exist within all modern nations. These differences have, of course, been part of government from ancient times. Athenian citizens were expected to vote, to serve as government officials if the lottery chose them, and to defend the city.[3] Paul got out of jail by identifying himself as a citizen of Rome, a special, protected class at the time.[4]

A key issue of government and citizenry is whom to include in the set of protected members of the community. Sometimes this is simply a question of subgroups and power, but it is often addressed as a question of purity in a similar way to religion, in which subgroups marked by gender, family history, ethnicity, or other factors are identified as "not real citizens" through negative purity perspectives.[5]

As with all of the group-definition processes we've looked at throughout this book, there is an important question of who qualifies as a full member of the community. Some definitions provide for breadth and inclusion, while other definitions are limiting and exclusionary. Of course, different subset definitions are more or less fair and include more or fewer individuals in the assurance game. A government that precludes who can be citizens or who can sit in the halls of power or that includes slavery for a subset of people is limiting who gets to play the assurance game and who is stuck in the sucker's deal of the prisoner's dilemma.

While government often includes formal definitions of roles within a society (citizen or not), it also often has aspirational goals that define the group. For example, most organized governmental systems have had flags for the group to rally around, whether it be the SPQR (*Senatus Populusque Romanus*, "the Senate and People of Rome") of the Roman republic held atop a standard for all to see, the heraldry of feudal houses in Europe or Japan, or the American flag lifted on the sands of Iwo Jima or still standing after the British shelling of Fort McHenry.

One of the interesting things about flags is that their actual structure is less important than the agreed-upon idea that the flag marks an identifiable group. Similarly, the specific design of money is less important than everyone agreeing that the money is a viable means of exchange. (The design of money to prevent counterfeiting is important to the trust that we put into it. Similarly, good flag design depends on our ability to detect visual signals from a distance.) While the specific designs are theoretically arbitrary, both money and flags provide important aspirational symbols for what a nation thinks is important. In 2020, the state of Mississippi finally changed the design of its flag from one centered on the Confederate battle design (symbolizing a rebellion in support of slavery) to one centered on a magnolia flower. US money tends to portray political figures, while other nations include scientists and other luminaries on their currency. The US uses postal stamps (a form of currency) as a means of honoring nonpolitical (as well as political) luminaries.

Flags often provide a symbol to rally around, but one can also rally around concepts and ideas. For example, many have argued that the real symbol of America is not the flag itself but rather the willingness to allow someone to burn that flag in peaceful protest. This brings us to the idea of first-party aspirational goals and the fact that governments are often built on defining mythologies (Athena laying down the law for Athens, Romulus and Remus and the self-construction of Rome, or the American Founding Fathers). One of the most interesting things about these myths is that they often outpace their own reality. This leads us to the idea that first-party aspirational goals play an important role in government.

FIRST-PARTY ASPIRATIONAL GOALS

As with religions, governments define a "greater good" that one is supposed to be working toward. Americans historically have had faith in the ideals set forward in the Declaration of Independence and the Constitution, both of which are treated as religious documents, pored over word-for-word by lawyers, like Talmudic scholars, trying to interpret the founders' meaning. From a certain perspective, we can see the Founding Fathers as American myths, from Washington's character to Jefferson's intelligence to Adams' integrity, with Benjamin Franklin as the older mentor shepherding the group to a new "republic, if you can keep it."[6] Moreover, we treat these physical documents as religious artifacts. The original documents remain on display at the National Archives, which is a major tourist destination in Washington, DC, where they are stored behind heavy glass in a grand room in a massive stone temple lined with columns.

Importantly, these aspirational goals often run ahead of the individuals writing them. The second paragraph of the US Declaration of Independence opens with the sentence "We hold these truths to be self-evident. That all men are created equal," but, of course, neither Thomas Jefferson nor most of the other men signing their names to the bottom of that document really meant *all* men. However, this document provided an aspirational goal that has been expanded and is now understood to include not only all men, regardless of race and caste, but also women and transgender and nonbinary persons.

Aspirational goals provide a guide for individuals making their own decisions, such as when judges argue over the implications and meaning of these documents or when laws are made to clarify the meaning of these sentiments. But aspirational goals also provide a means to argue and transform another person's point-of-view. Humans have the ability to reason through rule-based logic and to take the perspective of another person (through empathy and imagination). The combination of those two abilities makes it possible for one person to use those aspirational goals to change the mind of another. For example, Abraham Lincoln's journey to making the Emancipation Proclamation and supporting the Thirteenth Amendment depended in large part on Frederick Douglas's arguments of how the underlying aspirational goals of the US founding documents interacted with Douglas's actual lived experience. Similarly, Lyndon B. Johnson's journey to defending the Civil Rights and Voting Rights Acts depended on Martin Luther King Jr.'s arguments on the meaning of the Thirteenth Amendment. We can see a similar power in Ruth Bader Ginsberg's arguments that the Equal Protection Clause implied necessities of gender equality.[7]

STABILITY AND THIRD-PARTY PUNISHMENT

Of course, the vast majority of governmental rules are there to maintain order, to reduce intragroup conflict, and to ensure intragroup cooperation. Many governmental rules provide the basic necessities of coordination (which side of the road to drive on, the usable currency). But many governmental processes also provide polity-level infrastructure support for components that the group agrees are critical to the well-being of the community, such as education, insurance, communication, and health care. And many governments also provide for rule changes that can mitigate disagreements and shift people into an assurance game.

For example, one solution to preventing the selfish withholding of taxes in the tax game is to make it illegal and to punish people who do not pay their taxes. Similarly, we provide punishments for intragroup crimes such as theft, murder, and adultery. (Some of those punishments are explicit, such as jail time for theft or murder, while others change the consequences of intrasocietal interactions. For example, adultery shifts the balance of divorce proceedings, and negligence affects liability in civil suits.) Governments regulate relationships (such as marriage or the parent-child guardianships in adoption), enforce group-cohesive behaviors (such as vaccinations and schooling), and provide infrastructure that is cheaper to do as a society than as an individual (such as maintaining military defenses, providing resources for scientific discovery, building large infrastructure projects, and providing insurance, including fire, police, and medical services).

The question of how to enforce those rules, what punishments should get meted out, and what options for redemption exist are all moral questions defined by governments. Generally, governments apply third-party punishment in similar ways to what we described in terms of religion: rules should reduce the value of behaving selfishly, thus providing disincentives for behaving badly. An important element in both religion and in government is the shift from a dyadic retributive justice (*I will take revenge for what you've done to me*) to a group-driven justice (in religion, often stated as "God will punish you," whereas in government, criminal cases are always *Government v. Criminal*, not *Victim v. Criminal*). This is key to the idea of third-party punishment. By removing direct retribution for cheating (defection in the terms we used in the first part of the book), we can reduce the likelihood of a revenge cycle. The third party can step in, creating increased opportunity for restoring intragroup cohesion.

These community controls can be positive or negative. For example, they can create markets that enable the transformation of raw materials into amazing technologies. They can provide for regular elections and peaceful transfers of power, ensuring that disagreements will be reassessed at regular intervals. But they can also enshrine laws that separate groups (slavery enshrined in the original US Constitution, the Jim Crow laws of the twentieth century, the racial purity laws of Nazi Germany). They can also lock people in place, such as serfdom in the feudal era or sharecropping in the early twentieth century.

My point in this book is that those questions are all, fundamentally, moral questions. While the specific discussion of government regulations are beyond the scope of this book, we can see formalized government as an important tool that allows communities to grow larger. Government regulations shift incentives both at the monetary (fines, subsidies, punishments and rewards) and motivational (aspirational) levels. In addition, government can provide infrastructure and safety nets that can enhance productivity, provide opportunities, and reduce risk.

Our understanding of the science of morality will have consequences that we can study and appreciate. These laws have consequences because they change the games we play with each other. The argument at the start of the book is that we need to think about these laws in terms of bringing more people into the assurance game, making their lives safer, better, and more fulfilling. We can say that laws that tear people apart, that pit them against each other, or that enhance the power of one group over another are less moral than laws that bring people together. But if we build the laws right, we will find that we can increase both individual and community welfare. We can all do better when we all do better.

These decisions are particularly important as we look at questions of how to build institutional infrastructure. For example, the identification of who can vote and the distribution of how those votes are counted have enormous implications for power dynamics in a nation. (The original Three-Fifths clause in the US Constitution gave extra power to the slaveholding states since it counted three-fifths of the enslaved population but did not give them a vote. In that original constitutional construction, nonenslaved women and children were also counted as population but not given a vote. Similarly, the use of the electoral college in the US gives a disproportionate voting power to residents of smaller states.) Questions of how that infrastructure is to be organized are also important moral questions. For example, the US has been having a century-long discussion of which social services should be provided by governmental entities, how those social services should be organized, and to whom they should be delivered. (I suspect these arguments go back to the founding of the country, but they have been vocally argued since at least the New Deal and the 1930s.)

CORPORATIONS AND OTHER GROUP ENTITIES

We've looked at religion and governments as ways of organizing human societies. There are other codified attempts at group organizations, including corporations, nongovernmental entities, and other organizing principles that have similar properties. Two things make corporations different from the direct governmental entities described above.

First, interestingly, while corporations have an internal governing structure we can examine, corporations (like markets) depend on an external governmental structure within which they can define themselves. A "corporation" that built itself independent of any government would be a nation, not a corporation.

You can see this in the way that corporations choose which states and nations to incorporate themselves in to find the legal rules that will provide them the most advantage. While it is true that corporations have fought with governments (both legally and physically), their internal organization remains within some other governmental structure. For example, the various East and West India Companies (both the British and Dutch) had their own extranational armies and outposts controlled by company governors. They went to war with each other and with other nation-states.[8] The right to "build forts," "appoint governors," "declare war," and "provide police and justice" are written directly into the original charter for the Dutch West India Company. The British East India Company forced China to import opium at gunpoint. Both corporations maintained trading posts that were actually entire cities run by corporation governors. Similarly, the European Central Bank pressured Greece to pay its international debt, forced the government of Greece out of power, and required a new election to be held.[9] However, all of these multinational corporations wrote their internal legal rules within the context of a separate government (the British East and West India Companies within Britain, the Dutch within the Netherlands, and the European Central Bank within the European Union).

Second, the aspirational goal of corporations, to make money for their shareholders, is fundamentally different from most governments.[10] The first corporations were amortized insurance collectives. Trade via oceangoing ships in the 1600s was highly profitable but also extremely risky. While an individual ship might return very large profits, it might also sink or be taken by pirates, producing a loss of all investments. As with any insurance amortization, replacing an all-or-nothing investment in a single shipping venture with a small percentage of many trading ventures turned an all-or-nothing gamble into a highly safe, reliable profit.

What is important for our purposes here, however, is that aspirational goals matter. The purpose of the US Postal Service is to deliver letters across the country. This is why it costs the same amount to deliver a letter from Minneapolis to Washington, DC, as it does to deliver a letter from Minneapolis to Utqiagvik, Alaska (which requires mail travel by plane) or to the bottom of the Grand Canyon (Havasupai, Arizona, requiring a multihour trek down the canyon by mule). In fact, the original purpose of the US Postal Service was to increase intranational unity by increasing intranational communication.[11] Benjamin Franklin saw it as so important that it is explicitly identified in the US Constitution. In contrast, while corporations do provide services (airlines providing transportation, retailers providing goods, streaming services providing entertainment, internet service providers providing internet connectivity), these are all secondary goals. Corporations provide these services so they can make a profit. These services are how they make those profits, but the aspirational goal is not to provide the service—it is to make that profit.

The differences in aspirational goals between nations and corporations are important issues in discussions of privatization, particularly of services for which we, as a nation, have goals. Which specific goals are better served by private entities as a side effect and which goals are better served by public entities as a direct goal is a complex conversation beyond the scope of this book. But the ideas laid out in this book can help guide that discussion.

We can see lots of examples of institutions that define moral structures, including religions (whether built with a large infrastructure, like the Catholic Church, or not, like Rastafarianism), governments, corporations, and other entities. All of these institutions structure societies through definitions of in-group and out-group identification; identification of a higher purpose; aspirational goals that allow their members to play specific roles within that higher purpose; and structural rules that provide for sin, shame, and redemption as means to mitigate and mediate within-group conflict.

17 BEYOND HUMANITY

The question of community depends on our definition of humanity. Who is in that group of protected members?

Humans are not trying to solve the naive version of utilitarianism (share with everyone). Instead, we are playing the assurance game with a community of compatriots. This opens up the key question of *Who is included in that protected class that we are including in our assurance game?* If we were to define a group of people as not part of that class, then we would not have to play the assurance game with them and could, theoretically, treat them poorly, defect against them, and potentially kill them (this is the xenophobia side of parochial altruism). However, because the assurance game is not zero-sum we are better off bringing as many people as possible into that protected class so we have more people to cooperate with. In general, the progression of history over the last several millennia has been a trend to include more and more people in that protected "human" class. We have found ways to construct ever larger societies through the development of newer and better social and physical technologies. A key argument of this book is that this progression is an explicitly moral one; the more people we bring into that protected class, the better off we are as individuals.

One way to think about these changes is to consider human societies as increasing scales of the groups we live in. In thinking about these groups within groups, we can start from our *kith and kin,* our relatives (kin) and the small group of friends we interact closely with (kith)—the members of our family and our tribe. Neurophysiologically, we saw how oxytocin plays an important role in both identifying who is kith and kin and in defining those lines between them and the rest of the world. (Oxytocin is released when a mother holds her newborn child for the first time and between couples making out. It also sharpens the line of cooperation, increasing cooperation with others one considers close to oneself in a social distance metric but decreasing it for people outside that circle.[1])

Kith and kin are people we interact with often, making second-party interactions straightforward, although most such interactions are based on simple assumptions of cooperation (that is, first-party social controls) and are not based on pecuniary tit-for-tat negative interactions. (Kin, of course, also provide for genetic similarities, which can drive cooperation evolutionarily.) Both of these groups will have sufficient intragroup interactions to include third-party community controls as well should they be necessary. (Think of a parent scolding a child for hitting their sibling.)

Beyond kith and kin are the people we recognize as being within a larger, more general cooperative culture, perhaps defined by similar religions or similar ethnicities. Neurophysiologically, these are the people we can easily empathize with, which makes it easier to see their point of view and to cooperate with them in the assurance game. We can expand this group by building integrated societies and by increasing our experience with other perspectives.[2]

Beyond that are the people who are strangers but whom we recognize as human. These are people with whom we can negotiate; people with whom we can create legal constructions (religions, governments) that can ensure third-party social control. As we've seen, we can also use aspirational means (shared religious or governmental goals) to provide first-party social control within these larger and larger communities.

First interactions between highly different cultural groups often see the other culture as inhuman (beasts, untrustworthy, xenophobic). For example, the Romans and Etruscans were at each other's throats (as separate groups of strangers beyond kith and kin) until the Celts came pouring down out of the Alps led by Hannibal in the first Punic War. (Most descriptions suggest that the elephants Hannibal brought over the Alps were the key to the destruction of the Roman army, but it was actually the Celtic tribes that he was able to bring with him. Hannibal had Celtic-speaking lieutenants who convinced the Alpine tribes to join his army to go sack Rome rather than destroying Hannibal's army in the mountains.) The ethnic contrast of the Celts led the Romans and Etruscans to put aside their differences and fight together against the enemy at the gates.[3] One way in which communities expand is that we learn to work with others within our groups when other groups come over the border.

Importantly, all of these levels are fluid, and there are differences between where individuals draw the line, which has profound effects on their willingness to cooperate with members of broader and broader groups. As we've seen throughout the book, moral structures can change where we draw those lines and thus have an impact on our willingness to cooperate.

The now classic quote is to say that to unify humanity "we need an alien invasion" because that would unite competing human cultures against a new beast.[4] On the other hand, any alien capable of reaching us would, I am sure, be sentient (even if we were unable to communicate with them), and I suspect we would eventually translate them from strangers to community. That is, we would have to make the same transition that the descendants of the Romans and the Celts have made from competitors slaughtering each other into a joint within-group federation (the European Union). Hopefully, it wouldn't take two millennia to solve it. (The battle of Cannae where Hannibal destroyed the Roman army was 216 BCE. The formation of the European Union was in 1993, a difference of twenty-two hundred years.)

The story of humanity is a continuous shift to larger and larger communities. Early societies were made of kith and kin and competed (often violently) with other groups of kith and kin. But as communities grew, we found ways to work with larger and larger groups. The diverse modern societies we now live in include many cultures and ethnicities.

We have found ways to define communities that make diversity a strength rather than a burden. Consistently, communities that define themselves through acceptance of the other (as long as the other participates reliably in the assurance game) outperform communities that define themselves in contrast to the other. Compare, for example, the diverse Allies versus the other-limited Nazi Germany in World War II or the diverse Union versus the limited Southern Rebels in the American Civil War. Furthermore, because of the globalization of interactions that has occurred over the last five hundred years we now regularly interact with strangers from diverse cultures, which increases experience with, commerce between, and empathy across those cultures. We are now creating competition without conflict. We have many examples of human communities participating in nonviolent competition at one level and with shared cooperative multiple assurance games at another (such as sports leagues, within-nation federations, and cross-nation trade agreements).[5]

Here, though, I want to ask what happens as we move beyond humanity to animals, to robots, and to true aliens.

ANIMAL WELFARE

A moral question that has begun to be asked is whether human ideals of cooperation (often phrased in terms of *human rights*) should be extended to nonhuman animals. Humans have been cooperating (and thus playing an assurance game) with other animals for millennia (hunting dogs, rodent-hunting cats, war horses and war elephants, and domesticated farm animals). As far back as the Victorian era, if not earlier, humans began asking whether nonhuman animals had their own intrinsic rights to welfare.[6] As we increasingly understand nonhuman psychology, we can begin to ask how morality extends to nonhuman animals.

Over the last century, it has become clearer and clearer that a continuity of sentience exists, particularly across mammals. Other primates clearly have tribal politics and individual understandings of social structure, dogs and cats interact with humans socially, and we even have neurophysiological evidence that rats can plan cognitively.[7] Perhaps because of this continuity, we are often able to understand intuitively how other mammals are thinking and can often construct assurance games with them. On the other hand, the octopus is unquestionably intelligent but alien, in a way fundamentally different from us.[8]

I am not going to take a stand here on whether moral questions of rights should be extended to nonhuman animals. The point of this book is not to answer the question of what is moral and what is not but rather to argue that we have a new science of morality. Like all good sciences, this new science changes the questions we have to ask about our interactions with the world. Here, I merely point out that the animal rights questions look different when we ask what it means to "play the assurance game" with them, in part because it requires us to ask how they would want us to cooperate with them in the first place.

Pets

As an example test case, we can apply our understanding of the assurance game to questions of pets and their role in communities. Half of US households have pets. Forty percent of them have dogs, and 25 percent of them have cats.[9] If my students and colleagues are any indication, this number is rising dramatically in the current pandemic lockdowns we are living in. It is clear that many individuals, particularly in modern societies, include their pets in their concepts of family and community. (Humans' neural systems release oxytocin in response to hearing a purring cat or snuggling with a contented dog, which would help make the pet part of the family.[10])

Historically, pets have been seen as examples of hijacking our intrinsic goals to find family, to touch and be touched, and to share experiences with our friends. Pets tap into this human-human need. But I'm not so sure that keeping pets should be seen as a failure of the system so much as an extension of the intrinsic goals of community and family to nonhumans. The relationship between owner and pet, particularly (but not only) mammalian pets such as dogs and cats, is often clearly a mutual relationship between the human and nonhuman animals. (Remember, humans are animals too, therefore the common linguistic construction "humans and animals" is wrong. It should be "human and nonhuman animals.")

This idea that pets are not a misapplication of the need for kith but rather a created community relationship has important policy implications. Being a member of the community implies both rights and responsibilities. Communities survive by enforcing procommunity (cooperative) behavior and punishing anti-community (selfish) behavior. In the US, some animals are designated as service animals and provided rights that general pets are not. For example, service animals are allowed in places, such as restaurants, that general pets are not. On the other hand, service animals have responsibilities as well as rights.[11]

This issue of service animals versus pets has played out most recently in the airline industry, which is legally required to allow service animals on board planes. True service animals have been trained to perform specific tasks, and they are certified by a state or other governmental entity. (In the US, a service animal is defined in the American Disabilities Act as a dog [or a miniature horse] trained to perform a specific task for a specific person.) In contrast, the new term "emotional-support animal" has been created, which (to my knowledge) is not defined anywhere legally but is generally identified by a medical professional (you need a note from a doctor). The airlines have been trying to tackle the legal issues of what restrictions to put on emotional-support animals, which have ranged across the spectrum of available animals from typical dogs and cats to peacocks, pigs, and squirrels.[12]

While guidelines exist, to my knowledge there are currently no national or state laws on emotional-support animals, but we can ask how their interactions play out in our community structures. One issue is whether the nonhuman animals themselves have responsibilities that go along with their travel rights. Another issue is what responsibilities the humans traveling with those animals have to their fellow humans within the community. I travel a lot (I used to travel a lot before the pandemic) and have observed every sort of interaction, from cooperating (ensuring that their pet does not bother other passengers) to selfishism (for example, the person who got angry at their seatmate for insulting their "friend," a cat in a carrier, just because the seatmate was deathly allergic to cats. The flight attendant had to beg someone to switch seats with the allergic seatmate. Luckily there were cooperating people on the plane). Seeing these human and nonhuman animal interactions in the new light of community provides us a different perspective on how to incorporate them into our moral codes.

Climate Change and Biodiversity

The realm of climate change, as well as our interaction with the natural world, is the other major place where discussions of morality and animal welfare are taking place. Obviously, we are not going to try to play an assurance game with a polar bear or a redwood tree, but we are, in a sense, playing an assurance game with the earth itself. There is strong evidence that biodiversity is an important part of ecosystem productivity—that is, the more we maintain biodiversity in our world, the better off we are within it.[13] It is not that we need an individual polar bear to share food with us, but by maintaining that arctic ecosystem, we maintain our own.

It is important to recognize that even were we to shift the climate to kill off most of the species alive today and make the earth uninhabitable by humans, other animal species would continue to survive without us. The earth would still be here. The moral question is not whether we will destroy the earth— that's not going to happen. For all of our scientific prowess, we would never be able to shatter the planet or destroy the atmosphere. This isn't *Star Wars*, and we don't have a Death Star. Even all of the nuclear bombs in all of the nations of the world could never actually destroy the planet. On the other hand, it would be easy to make it uninhabitable by humans and really easy to destabilize modern civilization, which is so fundamentally dependent on specialization and efficiency.[14]

In the end, the question of climate change is the ultimate assurance game—we all share this pale-blue dot of Earth. And I suspect that we will need to use all of our moral tools to solve this problem, including negotiated agreements, third-party punishments, market forces, and aspirational goals. To me, one of the most iconic photographs of all time is *Earthrise*, taken by the astronauts on Apollo 8 as they entered lunar orbit. It shows Earth floating alone in space above the moon (a journey that required a lot of people to work together in the cooperate-cooperate corner of the assurance game).[15] It makes the limitations of this small planet we depend on abundantly clear. It makes clear that all of the cooperation and strife we have experienced over the millennia, all of our greatness and all of our failures, all reside on this pale-blue dot alone there in the darkness. It makes clear that we are all playing the same game together, the same assurance game. It makes clear that to succeed at this assurance game we will need to find a way to solve it, not only as competing tribes and groups but as one species, together, because in the end we will all succeed together or we will surely all fail together.

Over the last fifty years, our understanding of computation has been improving at a dramatic rate. It is now clear that the brain is an information-processing device that provides computations that allow us to interact with more and more complexity in the world. Note that when I say "information-processing device," I do not mean to imply that we are digital computers running software like your desktop or laptop computer. What I mean is that we process information about the world. (This is what it means to be "intelligent"—that we can take information about the world, combine it with our knowledge about the past [memory], and use it to make better choices about the future.) Given our current understanding of how brains implement intelligence and how that translates into "mind," it is no longer a question of whether we will be able to create an intelligent machine but rather a question of when we will create it.

This means that we are soon going to face the question of how to handle an interaction with a new being: AI (artificial intelligence), the intelligent machine. Questions of how humans will interact with these nonhuman machines have long filled the science fiction literature, from the early stories of helpful robots to intelligent ships endowed with a desire for self-preservation.

The term *robot* comes from Karel Čapek's 1921 play *Rossum's Universal Robots*, which is about how workers on a factory line become more automated and less human as their roles become more limited within the greater community. *Robot* comes from the Czech word *robota*, meaning forced work. The play is, in reality, about serfdom and slavery and the moral consequences of the dehumanization of early factory line work. While there are economic consequences of automation, the automated behavior of those robots (automated factory line machines) is not really sentient. The concerns of *Rossum's Universal Robots* are moral ones about the assurance game that we have already addressed. Instead, the new problem we need to worry about is that of *androids* (human simulations with individual desires and preferences).

Some of the classic android stories are from the Robot series by Isaac Asimov, which proposed that robots would require specific laws to interact appropriately with our community. In the Asimovian universe, as first laid out in his 1942 story "Runaround," robots must obey three laws: (1) a robot may not injure a human being or, through inaction, allow a human being to come to harm; (2) a robot must obey the orders given it by human beings except where such orders would conflict with the First Law; and (3) a robot must protect its own existence as long as such protection does not conflict with the First or Second Laws. Note that these are quite literally a moral code. (In the language we've derived in this book, they provide intrinsic goals and allow for first-person social control.) While many people have since interpreted the Asimovian laws as physical laws that robots would need to follow to work, that is clearly not true. One could easily build a robot that does not follow these laws. (Certainly, the robots from James Cameron's 1984 thriller *Terminator* and its sequels do not follow these laws.)

The Asimovian laws are definitions of how these robots (androids) should interact with humans. They are about how to integrate these nonhuman machines into the human community. As such, they are moral codes. Asimov's robot stories are, at their heart, explorations of moral philosophy. They are stories about how those moral codes play out in specific situations; for example, *Should a robot hold a human hostage if that human is set upon harming themself?* This was the premise of the movie *I, Robot*, which, while sold as an Asimovian story and containing the Asimovian title, more closely followed the plot of Jack Williamson's novella *With Folded Hands*, in which the human race ends up as bored dependents trapped in their homes to prevent them from encountering even mildly dangerous situations.

The idea that artificial life-forms (whether robots, androids, or some other general form of AI) will run rampant is an idea that is being hotly debated today. Some concerns include questions of overly simple AI motivations, such as the paper clip problem, in which one imagines an AI that is tasked with making paper clips and provided with sufficient abilities that it destroys the world trying to make paper clips out of everything. (Notice that what is going on here is that the AI has an intrinsic goal of making paper clips.)

Nick Bostrom eloquently discusses this issue in his book *Superintelligence*, in which he points out that this problem is larger than that of AI because we already have constructs that interact badly with our communities. For example, Bostrom argues that the corporation, which is tasked with making money for its shareholders and which contains armies of lawyers capable of working together over time toward this aspirational goal, is a form of dangerous nonhuman intelligence. We can see these corporations both as within-group communities (in which individuals are willing to work for the company as a group) and between-group interactions within a larger community (participating, for example, in lobbying for policies, providing advertisements and money to political campaigns, and making decisions that have implications for our physical climate and social environment). We will need moral control structures to address their interactions with the broader society.

Coming back to the question of individual robots (androids, artificial life-forms), we can ask how best to integrate these alien beings into our human communities. The Asimovian laws provide one potential moral code, but many others have been explored within science fiction, ranging from the synths in the British TV series *HUMⱯNS*, to the replicants of *Blade Runner* (based on Philip K. Dick's novel *Do Androids Dream of Electric Sheep?*), and the Cylons of *Battlestar Galactica* (particularly the 2004–2008 reboot).

In *HUMⱯNS*, the interaction is treated as one of racial tension in which individuals learn to recognize these new beings as human and bring them into the protected class of community members. Key to the story is the relationship between the Hawkins family and the synth Mia and their transition to seeing her as a member of their kith and kin to protect and defend. In large part this journey begins from demonstrations of empathy driven by Mia's mothering instinct (an intrinsic goal built into her) and the youngest Hawkins child, Sophie.

In *Blade Runner*, the replicants are provided a limited life span, forced to live off-world, and (theoretically) have no emotions, which makes them dangerous and unable to exist within a community of humans (without emotions, it is hard to empathize with another). The premise of both *Blade Runner* and the source novel *Do Androids Dream of Electric Sheep?* is the question of whether the hard-bitten detective Rick Deckard (a "blade runner" tasked with hunting down replicants hiding in the community) is himself able to exist within the community of humans. Replicants are identified through a test of Pavlovian emotional responses (see chapter 7) to emotionally laden images (such as violence to children or a calm beach and seascape). The deep question of both the book and the movie is whether Deckard has oversuppressed his emotions in his violent war against these human-appearing replicants and whether he himself would pass this emotion test. (These questions build on the idea that a soldier must sometimes give up their moral soul in order to accomplish the tasks of war, a concept we identified in chapter 11 as moral injury that is key to some forms of PTSD.)

In some versions of *Blade Runner*, particularly in the movies (see comments by Ridley Scott), there are suggestions that Deckard may be a replicant himself.[16] In the book, he is clearly human. In my view, Deckard as replicant destroys one of the main questions in *Blade Runner*—namely, whether a lack of emotion makes one less human or whether having dreams and desires is the key. (The title of the book, *Do Androids Dream of Electric Sheep?*, is a reference to Deckard's dreams of owning a live animal [see "Pets," above] and whether the replicants also have dreams and desires.) Throughout both the book and the movie, it is clear that the replicants are desperate to survive, to be part of a community, and to be allowed to come home.

In *Battlestar Galactica*, the interaction is one of conflict and war, addressing the topic of how both sides dehumanize the other, as is often done in war, and how reconciliation requires recognizing the humanity of each side. Fundamentally, all of these shows are asking how to integrate these now-intelligent machines into society, particularly given the fact that we would (hypothetically) understand their mechanisms.

This was most eloquently addressed in the famous *Star Trek: The Next Generation* episode "The Measure of a Man," in which the android Data (who was artificially created and thus is a machine with understandable mechanisms of decision-making) is threatened with disassembly. A computer scientist wants to take Data apart to attempt to replicate him. It is not clear that the scientist will be able to put Data back together again. Data objects to the disassembly, and the majority of the episode is a courtroom trial to determine whether Data has the right to choose his own fate. The two sides at first argue about the machinery of Data and the fact that he consists of understandable processes. But the episode turns in a powerful scene between Captain Picard (who has been tasked with defending Data) and Picard's old friend Guinan (played perfectly by Patrick Stewart and

Whoopi Goldberg, respectively), where they realize that the issue is not about Data's mechanism but about his role in the community. "In the history of the world, there have always been disposable creatures," says Guinan to her old friend. "You're talking about slavery," responds the shocked Picard.

When Picard returns to the trial, he defends Data not on the ineffability of his soul but rather on his interactions with the crew and his place in the community—his medals, the trust of his comrades, and, ultimately, the fact that he had a lover, who died, and the picture he keeps of her to remember her by. (One of my favorite moments in this episode is the judge's surprise at Data's intimate relationship with his former crewmate Tasha Yar. The judge looks at Commander Riker, another member of the crew tasked with prosecuting the trial, who looks down, embarrassed. That moment makes the reality of the relationship—the unkept secret within the community— real in a way that only human behavior can show.) As a member of the community, Data is acknowledged as fully human and gains the freedom to choose his own fate.

In the end morality is about finding a place within our community. And as our machines grow more intelligent, we will need to determine how to include them in our group dynamics as well.

IV THE NEW SCIENCE OF MORALITY

Morality is about mechanisms to solve the assurance game and allow us to work in coordinated groups, creating within-group cohesion and ensuring against selfish behavior. This implies that not all systems are equal. A science of morality does not imply moral relativism.

The big statement of this book is that what humans call *morality* is a set of mechanisms that help us turn our interactions into an assurance game and to find our way into the cooperate-cooperate corner of that assurance game, where we can reap the benefits of its non-zero-sum nature and make things better for all of us, both as a community and as individuals. Fundamentally, humans have evolved to work together. Morality is a set of tools that make us better team players.

The argument in this book can be summarized as a synthesis of the following three insights, each of which derive from a large literature of scientific study over the last many decades:

- *Assurance games are fundamentally different from the prisoner's dilemma.* In a prisoner's dilemma, what's best for the individual and what's best for the group are in conflict. But in an assurance game, it is individually better to cooperate if your compatriot is also going to cooperate.
- *People are not unitary decision-makers. How we ask the question changes the decisions people make.* The brain takes in information about the world (perception) and combines it with information stored within (memory) to make a decision (take an action). This means that decision-making

processes interact with the world one is living in, including the social world.

- *Changing the rules of the game can change a prisoner's dilemma into an assurance game.* These rule changes modify the social world one is living in and thus, through interactions with the decision-making systems in the brain, change the behaviors of the members of our community.

Bringing these literatures together, I have argued that we call the suite of social norms and institutional technologies that drive us into the cooperate-cooperate corner of the assurance game *morality*. This means that morality has an identifiable goal, and therefore some moral structures really are better than others. These moral technologies have real consequences on whether we make our societies better or worse places for the people living within them.

THE PROBLEM OF MORAL RELATIVISM

One of the biggest problems with studies of the mechanisms of morality is that they often fall into the trap of *moral relativism*, which is the idea that moral statements are culturally and contextually dependent, and thus one cannot say that one moral structure is better than another.

Much of moral relativism comes from the idea that science is amoral (amoral, meaning irrelevant to the question of morality, as compared to immoral, meaning against the moral goals we might have). This sentiment goes back to David Hume, who proposed a difference between *description* and *prescription*: description entails understanding how something works— *What are the causes? What are the mechanisms?*—while prescription entails what one should do—*What is the goal? What should we do to achieve it?* We have talked about this at several points in the book, including in chapters 1 and 12, where we noted that science often has prescriptive goals as well. Beyond the prescriptive goals of many sciences, however, the act of prescription itself is dependent on description—if you want to accomplish something, knowing the mechanisms of reality can help you succeed. The classic example is the observation that because bacteria evolve, antibiotics are useful for only a limited time, which means it is dangerous to use them excessively, so one should not use them to treat a viral infection. It is important to get the description right so that one can achieve the right prescription.

An anti-vaccination hysteria has gripped much of the US in the last few years, particularly in the false belief that vaccines cause autism. Vaccination is better for the group because diseases need a population of infected hosts to spread or they fade away, but if it really were true that vaccines cause autism, then one might hesitate before vaccinating one's child. That would turn vaccination into a prisoner's dilemma—it would be better for the group if everyone was vaccinated but more dangerous for any individual. However, vaccines do not cause autism. The original statement was completely made up in a fake study that has since been retracted; the lead author has lost his medical license and cannot publish in any respectable science journal ever again.[1] In truth, vaccination is safer for both the individual and the group. There is no dilemma here. Vaccination isn't even an assurance game. It's better for the individual to get vaccinated if they can. Full stop. It is also better for the community for as many people as possible to get vaccinated, but even if others in the community refuse to be vaccinated, it is better for you as an individual to get vaccinated.[2] Getting the science right is very important.

Furthermore, as we noted, science itself has prescriptive goals. In our definition of morality, the differences between prisoner's dilemmas and assurance games imply that descriptions of the consequences of social norms and institutional structures on behavior have prescriptive consequences, and some moral structures really are better than others. However, we need to be careful when we identify differences in the moral worth of different social structures.

BEYOND MORAL RELATIVISM

Historically, some arguments for moral relativism came in response to Western ideas of progress, which had suggested that larger, stronger, and "more developed" civilizations were inherently better than smaller, weaker, and "less developed" ones. The recognition that Western civilization was not necessarily better was probably a good response to the problematic assumptions of "progress" (that led to massacres, genocide, and slavery of cultures assumed to be "less far along"). On the other hand, this reaction has also led to a moral nihilism that says that we cannot judge anyone's culture, that we are reduced to describing what each culture does, and that all we can do is to measure what humans choose when faced with moral questions. (Contrast moral relativism as a solution to the cultural arrogance problem against an aspirational solution that says, "None of us have it perfect, but there are goals to strive for.")

In this book, I have taken the stand that the goal of morality is to create structures to turn our interactions into an assurance game. This hypothesis avoids the danger of moral relativism because it identifies a normative goal: moral structures that bring more people into an assurance game and make it more likely that we are playing an assurance game with each other are better structures. Because assurance games are not zero-sum, the more people we can bring into our cooperation group, the better our group will be. This means that these goals are not just about limiting our group to cooperators but also about increasing the size of that group by getting others to cooperate with us.

This idea has some similarity to that of *harm reduction* and has some components of *utilitarianism* and *collectivism*, but I have argued that there are fundamental differences between the idea of morality as a means to solve the assurance game and these three simpler ideas. Importantly, an assurance game is not self-sacrifice for a cooperative good. (That would be a prisoner's dilemma.) In an assurance game, the cooperative good is good for the individual player too. Similarly, an assurance game is not simple utilitarianism because the goal of an assurance game is to cooperate with those who will cooperate with you and not with those who won't. On the other hand, because the assurance game is a non-zero-sum game, the more people we can bring into the game, the better off we will be.

We saw that in many conditions an inherent cross-level conflict exists such that within the team, people who work for themselves first (we called them selfish) do better than people who work for the good of the team first (we called them cooperators) but that groups made up of cooperators do better than groups made up of selfishists. We saw that one can do things to encourage cooperation, including providing a sense of group definition, a sense of the value of one's role within the group, third-party punishment, reputation, gossip, shame, and the chance of redemption. We have seen how we can change rules to change games—tax penalties make paying your taxes good for you as well as good for the group. And we have seen how the creation of aspirational goals can change the definition of what we seek. This is one reason why simple monetary (pecuniary) rewards for good behavior or punishments for bad behavior can sometimes backfire, such as in the example of the day care that started to charge for late pickups rather than basing the punishment on social disapproval.[3]

Unfortunately, however, the assurance game actually has two stable conditions: one in which each individual puts a large portion (but not all) of their resources into the pot to create the infrastructure and community components and one in which the community collapses, and individuals retain most of their resources (or, more accurately, cooperation remains only within much smaller communities).[4]

THE BISTABLE EQUILIBRIUM

The assurance game implies that there are two stable end points to the question of cooperation. If everyone else around you is cooperating, then it is wise to cooperate with them as well, but if you are living in a society that is not a community, where you are "on your own," then you shouldn't be "cooperating" with those others.

Experimental studies of these games find that they do tend to bifurcate. They either approach a cooperative situation in which people trust each other and contribute to the community, or they approach an individualistic situation in which people do not trust each other and hold the money for themselves.[5] For example, in a tax game experiment with a multiplier of 1.6 (each $1 in the pot produces an extra 60¢ to be distributed back out), Urs Fischbacher, Simon Gächter, and Ernst Fehr found that most people (half of their subjects) contributed conditionally. These conditional cooperators were willing to contribute the same as everyone else, so the more the others contributed, the more they were willing to contribute.[6] This can create a virtuous cycle in which people contribute more as they see others contributing more. Similar virtuous cycles can be seen in trustee games and other assurance-like games. On the other hand, it can also create a vicious cycle when people see other people unwilling to contribute, which makes them less willing to contribute themselves, driving the average contribution down to zero. These vicious and virtuous cycles turn out to be culturally dependent, with some cultures even punishing people who share. The theory put forward in this book would predict that those vicious-cycle cultures are harder to live in and less able to reap non-zero-sum benefits. The experimental data are consistent with this prediction.[7]

This observation means that the assurance game story we have been exploring produces two competing visions of communities—cooperative communities where people contribute to the central pot and individual-istic communities where people hoard everything for themselves. So we might expect that the bifurcation will lead us either to communities that are not really communities at all (groups in which individuals are on their own, where they are basically reduced to hunting rabbits as individuals) or cooperative communities where people share together (everyone hunts deer together). This oversimplification is not what is actually observed.

For one thing, evolutionary simulations consistently find multiple niches within the community, many of which depend on the other mem-bers of that community. There are many situations where strategies can be successful if used by a small portion of the population but not if used by the entire population.[8]

For another, there is a big difference between providing a social safety net and communistic sharing. A social safety net is a form of insurance that allows someone to take a risk because if the venture fails, then they have not lost everything. For example, the development of the corporation was a form of social safety net that made possible dangerous trading ventures. On the other hand, there needs to be some reward for doing the additional work.

This continuum was best described by the infamous Laffer curve, which gained popularity as a justification for lowering taxes in the 1980s. The Laf-fer curve is a statement that is both trivial and profound (as many are). Laffer pointed out that with 0 percent taxes, there is no government revenue in the central pot, but with 100 percent taxes, no one has any money to spend, so there is also no revenue to put into the central pot.[9] Logically, this implies that there is a point that maximizes government revenue, with negative falls on either side. Because of his politics, Laffer and his colleagues used this to argue that lowering taxes would increase government revenue. But this is an empirical question: Are we above the peak or below the peak? If we are above the peak, then, yes, lowering taxes would increase revenue. But if we are below the peak, raising taxes increases revenue and what's actually available in that central pot. Mathias Trabandt and Harald Uhlig empirically measured the Laffer curve by fitting tax revenues over a number of countries and found that the peak lies somewhere around 70 percent average taxes. Every country was below the peak. The only country they found that was even close to the peak was Sweden.

Sociological studies have found large differences in these cooperation decisions between societies that can be identified as *social capital*. Some communities contain social norms and infrastructure that tie the community together, while others are fractured apart. For decades now, social scientists have been concerned about decreases in the social capital within the US and the consequences of those changes.[10]

From as early as Ibn Khaldun (the Tunisian scholar from the Arab renaissance, from whom we got the term *asabiya*), people have suggested a generational cycle of empires in which a collaborative group comes in and takes over due to the ability of the members to work together (asabiya), but then internal strife appears as selfish individuals (and subgroups) compete for power within the new community, leaving it vulnerable to the next dynasty. This cycle of collaboration and collapse should not be taken as inevitable but rather as a commonly observed trope within history. (History, as is noted in the quote often attributed to Mark Twain, doesn't repeat but does rhyme.)

However, it is important to note that these discussions are usually less about true individualism (living as a hermit, alone) and more about whom to cooperate with. In communities with low social capital, people tend to band together in small groups of families, churches, and neighbors. We can see this dichotomy in the famous Margaret Thatcher quote in which she said that "there is no such thing as society. There are individual men and women and their families. . . . It is our duty to look after ourselves, and then, also, to look after our neighbors."[11] Thatcher was arguing for individualism and the destruction of the state as a safety net. But notice that what she ended up actually arguing for was a small, localized safety net made up of families and neighbors.

Part of the question is whether people want to share beyond their small family groups and whether they recognize that the larger groups have, in fact, been cooperating with them. Many of the sentiments that break communities apart lie in the belief that some individuals are getting more than their fair share and that free riders and defectors are taking from their community. In part, I suspect that some of the issue here is that people don't recognize that the help they got was actually help, as evidenced so eloquently by the clearly unrecognized irony of the quote from actor Craig Nelson: "We're a capitalistic society. I go into business, I don't make it, I go bankrupt. They're not going to bail me out. I've been on food stamps and welfare, did anybody help me out? No."[12] Of course, not only are food stamps and welfare obviously governmental help that others in the society have paid for, but the legal definition of bankruptcy, which allows one to rebuild one's monetary and credit abilities in the face of crushing debt, is also a societal structure for bailing people out.

Recognizing that others in the society have helped you when you stumbled is likely to be a vital part of the construction of successful societies. Similarly, so is recognizing that the help you are providing for another is aiding them in getting back on their feet. Because of the non-zero-sum nature of the games we're playing, successfully recovering that cooperative relationship is going to be better for both parties. Morality provides tools to help us navigate those social interactions. We can take the knowledge that we've worked out over this book and identify some key points to creating better communities where more of us get to participate in a shared assurance game. In the final chapters, we'll summarize some of those points that we can derive from the logic so far.

19 BETTER MORAL STRUCTURES

We can use the knowledge derived from moral experiments and from our understanding of decision-making systems to identify the factors that go into a positive moral structure capable of bringing more and more people into the (non-zero-sum) cooperation corner of the assurance game.

A good moral structure helps bring everyone in the community into the cooperate-cooperate corner of the assurance game. Because the assurance game is non-zero-sum, the more people we can bring into our assurance game, the better off we will be. As noted in our initial discussion of the assurance game in chapter 3, three problems can push us out of an assurance game: that of the free rider, who lets everyone else cooperate but takes selfishly for their own; that of the enemy at the gates, where another (external) group comes to compete with our group; and that of the enemy within, where a subgroup is willing to cooperate within their own subcommunity but unwilling to share beyond that. For each of these, we will need to create social structures that can handle both the con artist (who lies and cheats) and the bully (who uses violence and force).

Here, we'll examine each of these issues and identify their consequences for groups, both in terms of group cohesion and prevention of within-group takeover by selfish parties. We'll also examine these issues in multilevel groups, in particular how to achieve competition without violent conflict. One of the most useful things we'll find will be interrelated networks of groups such that many paths exist to find connections between individuals.[1] However, as noted throughout the second half of the book, these should not be taken as a recipe for a successful community but only as starting points arising from the logic of the first half of the book. There is still a lot of science to do here.

That being said, the following ideas can encourage cooperative behavior:

- *A definition of the group*—membership in the group needs to be clear.
- *Being valued within the group*—a role within the group gives us a purpose.
- *Third-party punishment*—being selfish within the group should have a personal cost.
- *Opportunities for future positive interactions*—ensuring that there are paths to cooperation and redemption prevents one from cutting off future opportunities.
- *Aspirational goals*—cooperative goals that members of the group believe in can provide internal arguments to support first-party self-control.

WE NEED GROUPS

The mathematics of morality derive from tension and conflict between within-group and between-group dynamics. Within the group, selfish behaviors win out. While there are mechanisms (third-party punishment, gossip, shame, and redemption) that can reduce selfish behaviors, those mechanisms are not foolproof. In the end, selfish behaviors will often find a way to do better within the group than cooperative (altruistic) behaviors. However, between groups, cooperating groups win out over selfish groups. This means that over time selfish groups will diminish and fade, while cooperating groups will grow and spawn. The key factor is to ensure that the group can grow and fade without hurting the individuals within it; for example, we can ensure that it is possible (and safe) for humans to switch groups. This will allow humans to migrate toward groups that are cooperating and flee groups that are selfish. Intergroup dynamics don't have to be violent. There are ways to create intergroup competition without violence (see "Groups within Groups," below). But the ability to switch groups is important.

Between group competition is not only good but necessary. The ability to switch allegiances is a key to group competition.

DEFINING THE GROUP

We can define a group in many ways, but a few important markers stand out. First and foremost, the group must be identifiable. One must not only be able to recognize oneself within the group but be able to recognize others within the group.

One way to define the group is in contrast to the "other." We have all seen how one can define oneself as part of a group by defining what one is not. In the 1930s and 1940s, the German Nazi Party unified a subset of their populace by defining a non-Aryan "other." In the American South, during slavery and the Jim Crow era, racist powers maintained their voting block because they convinced the community to define groups by race, convincing poor White voters to see themselves as aligned with rich White plantation owners against Black communities. But defining a group this way has dangers and negative consequences.

Defining a group in opposition to an "other" means that your group cannot include talented people who are in that "other." A group defined by the "other" is leaving opportunities on the table. Many of the critical scientific discoveries that helped win World War II were made by refugees who had fled Nazi Germany because they had been declared "not part of the group." Similarly, the Union won the American Civil War in large part because the South depended on slavery, and many slaves fought back, either through active sabotage or joining the Union army to fight as soldiers or just through leaving the fields as refugees.

In contrast, there are other ways to define a group. Many groups are open to all but define themselves through difficult entry thresholds—a test of adulthood or a complex entry routine. By making entry difficult, you enhance the value of the group, in part because people judge the value of things based on the costs spent to acquire them but also because the selfish tend to be unwilling to spend a lot to achieve their goals.[2] Therefore, groups with difficult entry requirements tend to keep out the people who aren't willing to play along.

Moreover, these groups become more tightly knit when members perform these difficult entry tests together—think of the cohesion of elite fighting forces that have trained together. Being put through a shared trial produces group cohesion. Continued shared rituals are a key part of helping form and maintain groups. Humans are cognitively structured to derive group cohesion from participating in rituals together, singing, dancing, being in rhythm together.[3] I bet that a fractured group could be repaired through the context of shared rituals.

Inclusive groups that allow anyone to join will (in general) outcompete exclusive groups that define themselves in contrast to an "other." Moreover, inclusive groups that provide more opportunities for individuals will (in general) outcompete inclusive groups that are stifling. In the Cold War, many more people defected from the autocratic Soviet Union to the more democratic West than from the West to the Soviet Union.[4] (Some people did defect from the West to the Soviet Union in the Cold War. The reasons they gave were generally based on beliefs that the income equality ideals of the Soviet Union were more fair than the income inequality of the West, making it clear that the reasons for defecting were still moral comparisons of groups and teams and fairness. Note that switching sides here is even called "defecting"—they were rejecting the within-group cooperation of their own team for the other team. Of course, they were cooperating with the other team because they thought they would be more likely to find themselves in the cooperate-cooperate corner of the assurance game on the other team.) How a group defines its internal moral code will change who wants to join it to cooperate in its assurance game.

If we allow humans to switch groups, then the rules within a group matter. Inclusive groups will generally be stronger than exclusive groups.

Competition without Conflict: Groups within Groups

Many books, particularly those building from the anthropological and ethological literatures, have identified that humans show parochial altruism, cooperation within the group but xenophobia between groups.[5] It is certainly true that intergroup competition has historically often been extremely violent; however, the assumption that intergroup competition must be violent is false. (Remember, the goal is the cooperation; violent intergroup competition is just one tool with which to increase cooperation within the group—and it's not a particularly good one.) Modern humans live in nested levels of groups. We have family, relatives, neighbors, towns, states, countries, alliances. It has almost certainly always been so, as tribal groups identify both with families and moieties within a working tribe, as well as with larger, cultural groups.[6]

Importantly, one does not have to create xenophobia between groups to have group competition. Group competition can occur within a metalevel group. For example, sports leagues consist of teams, and these teams compete with each other. As teams develop new technologies, they can do better than teams that don't. A good recent example is the development of making baseball management decisions structured on individual contributions to team success (getting people on base, getting them around to score) rather than characterizations of individual mechanics. The way this transition occurred through team competition within the metagroup of Major League Baseball (MLB) was popularized in the book (and movie) *Moneyball*.

There is no question that teams that play well together (that have asabiya) are more likely to win sports games against teams that don't. But everyone in the National Football League (NFL) or in the National Basketball Association (NBA) or in MLB see themselves as part of a larger group. Watch NFL players hugging the other team after the game. Olympic athletes compete with each other, but they also see themselves as part of a sport. One of the most touching moments I've ever seen was when Norwegian cross-country skier Bjørn Dæhlie insisted on waiting twenty minutes for the last person (in this case Kenyan skier Philip Boit) to finish the race before going to the ceremony to receive his gold medal (in a race in which the difference between the gold and silver medals was eight seconds). The denouement of the wonderful movie *Cool Runnings* is when the East German team finally recognizes the Jamaican team's mettle and welcomes them into the community of Olympic bobsledders, looking forward to competing with their subgroup in future games.

Cool Runnings dramatizes a set of racial, economic, and ethnic disparities. In the movie, the East German team's unwillingness to see the Jamaican team as serious contenders reflects these unfair issues, providing a foil for the Jamaican team to prove themselves against. In reality, the real Jamaican bobsledders said that the other teams were supportive and encouraging and welcomed them into their inclusive group of those who think sledding down a winding, icy track at a hundred miles per hour is a good idea.[7] Bringing the Jamaican team into the fold was a popular move and increased people's interest in bobsledding. By being an inclusive group, the bobsled community was able to expand the success of its sport. Of course, there are many famous cases where sports groups were less willing to accept new players (e.g., baseball and Jackie Robinson), but as those communities expanded the pool of talent to draw from, the game improved.

Groups can compete through cooperation in a metagroup. These intergroup competitions do not have to be violent.

Intergroup Networks

We live in interrelated networks of groups. Individually, the groups we belong to are not set into a rigid system. While it may be true that there are political levels of town, county, state, country, and so on, many other groups exist with separate systems of their own. For example, many people identify as members of a religion, which often has expanding circles of membership, from one's local house of worship to broader communities within the religion to shared beliefs with other religions. Sports fans may root for a specific team but generally also prefer one sport over another. And this is not even taking into account the complex interactive set of unofficial groups we can belong to—work groups, hobby groups, reading groups, and even just groups of friends.

The more tangled a web of groups we live in, the less likely we are to be in specific conflict because a tangled web provides more ways to connect with each other. Of course, there are examples in which conflict over one connection can split the cooperation across another (such as the infamous cases of brother against brother in the American Civil War or the religious wars in Europe in the 1500s). However, fascinatingly, when there is a mismatch such that one connection leads to cooperation, while another leads to conflict, the cooperative connection tends to reduce the conflict connection more than the other way around. Many examples exist of communities in conflict that have been brought together by a shared relationship, such as a shared love of a sport to play together, a family relationship, or identifications of shared culture. Even across large-scale groups such as nations, the more nongovernmental organizations there are that span a pair of nations, the less likely they are to go to war.[8]

One of the new scientific questions that our science of morality identifies is how to build, maintain, and use these interwoven webs that can increase cooperation and reduce conflict.

A tangled web of groups provides more avenues for cooperation and reduces conflict.

Loners and Subgroups That Don't Want to Play

As we've noted throughout this book, the advantage of these assurance games is the non-zero-sum nature of the payout matrix. In an assurance game, if I am going to cooperate, then it is better for you to cooperate as well (in contrast to the prisoner's dilemma and other games). In this book so far, I've addressed these issues in terms of people who are members of a group but are not cooperating within it (individual free riders, subgroups that want to take from the larger group but not contribute back). This raises the question of what to do about someone who simply doesn't want to be part of a group. What about the loner? Or the subgroup that wants to go live out on their own? Do we need to treat them as an enemy at the gates?

The enemy at the gates is an opposing force that is trying to compel you into the negative side of the payout matrix. It is not someone who simply doesn't want to participate in your group. While the parochial altruism that underlies human behavior often treats outsiders as enemies, it does not have to be so. Because cooperation is better, the more logical decision would be to leave them alone while trying to reason with them and find a way to reach an equitable group interaction.

On the other hand, when dealing with subgroups and alternate communities, we need to be aware of whether those communities are playing assurance games within themselves. I would not want to argue that it would have been acceptable to allow the Southern states to break away and maintain slavery. (This violates one of the key tenets we included above—that individuals must be allowed to switch groups.)

Furthermore, we need to recognize that it is harder and harder to truly be on one's own, neither contributing positively nor negatively to the global community. In my house in Minnesota today, I am smelling smoke from wildfires in California. It would not be morally right for someone to say they wanted to be on their own and then pollute shared lands and resources.

As we noted earlier, many people think they are doing these things on their own when they are actually part of the intergroup network. Helping people see that interaction can be a way to bring them into the fold. I don't have easy solutions for such questions. The process is going to be tricky and non-trivial. Nevertheless, it is better to keep options open than to close them down.

Moral codes that leave open future opportunities for cooperation are better than those that permanently close them down.

INCLUSIVITY

The most important point made throughout this book is that the assurance game is not a zero-sum game. If we are playing an assurance game, then I, personally, do better if we cooperate. In the tax game, the more people I can get to contribute to the central pot, the better I, personally, do. (The more money that goes into the central pot, the more there is to pay out after the pot doubles.) Of course, these games are metaphors (simulations) for real situations where working together leads to a better outcome than working alone. We've seen how insurance spreads risk around (whether it be ships traversing dangerous sea routes or fire departments protecting our houses), how hunting together allowed early groups to gather larger game, and how central-built infrastructure is more efficient than individual constructions. The key idea is that in these non-zero-sum games, your gain is not my loss.

Interestingly, physical technologies can also provide for increased non-zero-sum gains from an assurance game. Innovation builds on innovation. Faster communication provides for better communication and better coordination between the members of our group. There's good evidence that the success of larger and larger communities has arisen in large part because of improvements in communication.[9] Coordination is an important part of maintaining group dynamics, but other social factors are important as well.

Roles

A key factor in group success is knowing one's role within the group. As we've seen, human groups are not just collections of individuals but collections of individuals playing specialized roles. I'm a firm believer in specialization—while I do believe that I could manage to learn to renovate my kitchen or fix my roof, a contractor who has spent a decade becoming an expert is likely to do a better job. Similarly, I can contribute to the good of the group via my specialization (making discoveries) in ways that someone who has not spent decades studying computer science, neuroscience, and cognitive psychology could not.

In early group definitions, roles were often preordained, defined by sex, race, or family history (caste). But nothing in our group definition requires this. In fact, one can make the case that a group that allows flexibility in role definition is going to do better than one that restricts it because, while there may be correlations between individual factors and the ability to perform a role, correlations are only probabilities. Furthermore, those correlations tend to be very small and tend to be swamped by the diversity within a population.[10] For example, men are, on average, taller than women, but the distribution of heights is so large that many women are taller than many men. Furthermore, height is affected by a lot more than genetics. The effect of nutrition on height swamps any effect of the genetics of sex. So even if some role actually depended on being tall (or short), one wouldn't want to limit that role by sex, race, or caste, even if one did limit it by height.

A group that rejects an individual's preferred role is inefficient (people are more willing to push for their self-driven passion than for the job they were told to do), wasteful (because that is likely to be a talented individual capable of accomplishing those tasks), and likely to drive an individual away to another group, particularly if another group offers them the opportunity to pursue their passion.

Nevertheless, a group should provide a means for an individual to find their role within the group. We should expect successful groups to provide opportunities for exploration and education and should recognize the importance of all roles within the group. We want to identify roles for ourselves within a higher purpose. We see this in any human community, whether it be families, groups of friends, communities of work, military operations, government, or religion. In any system where humans form a community, we look for roles within that community. And it is important that those roles are respected.

Moreover, there's no need to assume that a role, once chosen, must be fixed for life. By providing opportunities for change, we can enable translational skill sets and provide opportunities for individual discovery. A key thing seen within many successful communities is that they provide opportunities for advancement within that community—as one gains knowledge and skills, one takes on different responsibilities. But in a stable and solid community, every role is important, and every member of the team feels they are contributing to the success of the team.

Having a purpose in the group is an important factor in one's willingness to be a team player within the group.

FAIRNESS OF REWARDS

We have seen throughout our discussion that humans have an intrinsic goal of fairness—we want each of us to have "our fair share." But, for most people at least, the definition of "fair share" is more complicated than "equal." No society has succeeded at providing true equality between all subjects (someone always ends up "more equal than others"), and it is not clear that humans desire that level of equality. When Michael Norton and Dan Ariely surveyed Americans on the levels of inequality they wanted to see, subjects argued for a low but nonzero level of inequality.[11] Interestingly, these levels are typical of those nations consistently rated as the best places to live in the modern world (like Denmark).

Economically, humans generally believe that working harder should provide more rewards. We saw this in the ultimatum game. If Player A thinks they have worked for their $20 stake rather than receiving it from the experimenter, they share less. In fact, if Player B thinks Player A has worked for their stake, they accept lower offers. This can even be seen in children, who argue that fairness extends both in terms of work and rewards. Rewards are a way to ensure that we share the work.[12] (Note that sharing the work is a form of cooperating—someone who does not share in the work is being selfish.)

An interesting issue is whether equality should lie in outcome or opportunity. When pressed, most people actually seem to believe in equality of opportunity more than equality of outcome. Generally, the belief seems to be that we need to provide some encouragement for work, which need not be very high in order to provide incentives. Taxes can be very high as long as increased work always brings increased resources. We also do not want to produce flat taxes across the board. Taking 70 percent of one's income when one has billions leaves one with more than enough money to enjoy a spectacularly good living, but taking 70 percent of one's income when one has just enough to live on is deeply problematic.

There is an important danger in discussing equality of opportunity, which is that humans are individuals, with individual talents and individual preferences. We want to make sure that we reward people playing all the roles within our society. It is important to recognize that someone doing a job that anyone can do frees up the person who can only do the specialized job. And we should compensate them appropriately. Respect for all of the jobs in a society is an important part of building a successful community.

While we may say that some jobs are more important than others and modern economics provides more resources to some individuals over others, it is not at all clear that current distributions are made on the necessity of a role played within the community. For example, in modern societies many of the resource distributions are made based on the scarcity of jobs, as well as on historical reasons of racism and sexism, rather than importance. (Why are hedge fund managers paid so much more than teachers? That's insane.) In the US, as jobs shifted from being male dominated to female dominated, salaries decreased.[13] In the current pandemic, it has become clear that the actual "essential" jobs are generally paid lower wages (meatpacking industry workers, grocery-store clerks, delivery personnel, garbage collectors, teachers) rather than higher wages (lawyers, hedge fund managers, advertising executives). We will have to see whether the consequences of this recognition will change the pay structure at all, but I doubt we will start paying teachers like hedge fund managers, even if we should.

Opportunity, respect, and compensation for all roles is an important part of the moral code.

One of the main purposes of human communities is to provide safety nets. In a sense, a safety net is the ultimate example of the assurance game—what do you do when your neighbor is in trouble and needs help? Being part of a group means that you can trust that just as you have their back in their time of trouble, they will have yours in your time of trouble. Moreover, safety nets increase fairness.[14] Providing a safety net for people who are really down can make a huge difference in enabling them to recover and to return to society.

One fascinating issue of safety nets is the question of free riders. Over the last fifty years a long discussion has taken place on whether safety nets should include mechanisms to ensure that free riders cannot take advantage of them. The question of whether a safety net should be tested for free riders is a complex and nontrivial one that requires economic experimentation. In any complex system (and the societal nature of these economic questions is definitely complex), there are side effects and nontrivial consequences of these decisions.

Means testing is expensive. Keeping track of which students come from families that should get a free breakfast in schools costs money. My kids' school district found that simply providing a free breakfast to any student who wanted it was actually cheaper than keeping track of it, so they stopped means testing and offered healthy breakfast options the kids could pick up on their way to class. They found that this had the additional benefit of making it impossible for the kids to know who needed the breakfast and who didn't. Because the majority of kids are always hungry, there were always lots of teenagers eating that breakfast, which made the kids who were food-insecure at home more willing to take the breakfast, which allowed them to concentrate better, behave better in class, study more, and learn more. Removing the means testing actually made the system better all around.

Furthermore, limiting free things to those who cannot afford them changes the distribution of who wants to fight for it. Public libraries in the US loan books to everyone, whether they can afford to buy those books or not. By including both the poor (who cannot buy books) and the rich (who could) as potential library patrons, libraries create a base of support that includes an entire community, including the politically more powerful economic elites.

And finally, there are reasons that we, as a community, might want everyone to have access to these safety nets. We don't necessarily want everyone to make their own bet on their risk. For example, diseases need a population of hosts in order to spread. If someone is sick with a communicable disease, it is better to have that person get tested, treated, and stay home than to have them go to work and infect their coworkers. By not including sick leave in our society, we force someone who cannot afford to stay home from work to choose between losing money by staying home (potentially even losing their job) and getting paid to go to work (potentially infecting others). This turns the decision into a form of prisoner's dilemma—going to work so as not to lose their job is better for the individual in this scenario but could spread the communicable disease, which is worse for the group. Ensuring that everyone has access to sick leave and reasonable medical treatment is, in fact, a boon to all of us, beyond the shared costs. Providing sick leave turns this into an assurance game: better for the individual (stay home, get well) and for the group (stay home, don't infect anyone else).

Safety nets put a floor on how low one can end up. The safety net is the equality of outcome that allows us to provide an equality of opportunity.

PUNISHMENT AND REDEMPTION

In the cross-level tension we've been discussing throughout this book, one of the most important factors is the mechanism within the group to punish selfish players. We've seen examples of third-party punishment, whereby an individual who cheats is punished not by the person they cheated but rather by the group itself. This group-derived third-party punishment has four important effects. First, by deriving the punishment from the group rather than the individual, it reduces the responsibility of the individual, shifting the individual decision from one of whether to punish or not to a decision of whether to be a part of a state that punishes or not. This has the second effect of reducing the potential for retributive feedback. Third, it shifts the punishment from an emotional, reactive response to a planned, preset response. Second-party punishment is Pavlovian; third-party punishment is deliberative. In a sense, setting the punishment response as a defined legal consequence administered by a third party is a form of precommitment—each member of the community agrees to abide by the consequences of making an illegal decision. Finally, ensuring that a community has third-party punishment can be a form of *signaling*, providing information to community members (and potential community members) about the expectations within that community. This could increase a cooperator's willingness to join such a community, in part because they know that it would decrease a selfishist's willingness to join such a community.

A key to the success of groups is graded punishment for crimes, whereby small crimes provide small punishments that allow a journey back to society, and repeated crimes increase in punishment, making the journey back harder with each repeated refusal.

As we noted in our discussion of justice (chapter 14), punishment has two goals—redemption of the individual and prevention of the crime in the first place. The first depends on assumptions about learning and interacts with the different decision systems, each of which learns in different ways (see chapter 7). The second is an economic assumption and depends on assuming that the decision to commit the crime takes into account the costs, so increasing those costs can decrease the behavior.

We noted that institutions (governments, religions, social norms) can provide for the restoration of loss through compensation and return, making it possible for individuals to coax trust back through subsequent behaviors. And we noted that questions of redemption often depend on acknowledgments of wrongdoing, understandings of motivation, and forgiveness. Issues of empathy and understanding are important steps toward this question, as are issues of reputation and communication. An important issue as groups get larger and larger is that the number of times one interacts with a given individual decreases. This is a simple statement of mathematics. As groups get larger, it becomes important to create mechanisms of reputation that can carry information of interactions across those communities.

Punishment must be perceived as fair for a group to be successful. I think one of the best discussions of this lies in the arguments of the philosopher John Rawls, who argued that the moral statement of laws should be laid down ignorant of your position in society. Thus, if you believe that the king should be above the law, you don't get to assume that you'll be king. If you are okay with the king being above the law and taking whatever he desires from the peasants, you have to recognize that you may end up being a peasant in this society. These ideas lead to societies with aspirational goals of fairness and equity.

Moreover, an important step to redemption is the first-party self-control that creates an internal rejection of negative interactions: "I'm not the kind of person who does that." Current theories suggest that humans evolved mechanisms to understand the complex behaviors of the other humans they were interacting with and then applied those mechanisms to their own behaviors.[15] This tool kit can be applied to the moral realm as well. This tool kit enables first-person self-control, but the mismatch of one's behavior and these aspirational goals can lead to moral injury. Redemption needs to include a path back not only to society but also to one's own internal view of oneself.

Punishment should be aligned to decision-making systems and should be graded, allowing for a process of redemption. Aspirational goals can help prevent negative interactions through first-party self-control.

ASPIRATIONAL GOALS

This last point brings us to the important observation that successful groups usually have aspirational goals and, importantly, that those aspirational goals matter. Many corporations make their decisions under the assumption that their goal is to make money for their shareholders. In contrast, governmental entities have the goal of accomplishing certain tasks. For example, the goal of the US Postal Service is to deliver the mail. While recent governmental regulations have demanded that they make a profit doing so, the official goal of the post office remains to deliver the mail. This means, for example, that the post office should (and does) prioritize reliable delivery to everyone in the US, no matter how remote they are, over profitability. Similar arguments have been made about health care delivery, differentiating between national health care systems, such as the UK's National Health Service, and the private health care insurers that form the basis of the US health care system. The UK's National Health Service has the aspirational goal of improving the health of its citizenry (while keeping costs reasonable). The private insurers in the US have the aspirational goal of making a profit (while delivering health care to their clients).

Aspirational goals provide a moral force, whether they are religious ("treat your neighbor as yourself"), cultural ("honor your ancestors"), or national ("all men are created equal"). These goals have two effects on communities.

First, aspirational goals drive individuals to maintain their own behavior within the context of those goals. Humans are narrative creatures. The story we tell ourselves about who we are is an important part of what drives our actions. That story depends on those aspirational goals: "I want to be someone who . . ." As such, aspirational goals can drive morally positive decisions even when no one is watching because to take the other (morally negative) action would be incompatible with who we want to be.

And second, aspirational goals provide an argument for others to act, whether through second-party dyadic control (allowing one to say, "You know that this is wrong") or through third-party arguments of "Are you going to let this happen here?" It gives us an opportunity to make the argument to someone conflicted between a morally positive decision and a morally negative one (such as the pressure that can be put on an authoritarian regime by peaceful protests, which work surprisingly well[16]). It gives us an opportunity to argue that a stronger third party should step in when someone is being oppressed.

Of course, not all aspirational goals are good. The "Final Solution" of the Nazi Third Reich to exterminate the Jews whom they didn't want in their exclusive group was also an aspirational goal and led to more than 6 million dead. It is a tenet of this book that genocide as an aspirational goal is prescriptively morally bad because it removes community members from participating in the assurance game. (As we noted, one of the major reasons the Nazis lost World War II was because the people they had rejected from their assurance game group left to join the other side and came back to kick their a**.) As cultures grow and change, as we gain empathy with others and see our communities as becoming more inclusive, incompatibilities between aspirational goals can become apparent. In the US, the aspirational goal of "Manifest Destiny" that led to wars of conquest and to great suffering of shifted native populations is no longer seen in a positive light. In fact, as we noted, the post–World War II German governments have apologized and continue to this day to pay reparations to survivors and to some of the damaged communities.

Similar ideas of reparations are making their ways around the world.[17] The most appropriate process to provide justice remains under debate, but the idea that we can recognize that choices were wrong, crimes were committed, and redemption is required depends on that aspirational goal of creating better communities that are more inclusive and more fair. The question of how to balance retributive, restorative, and redemptive justice and how to construct better and more inclusive communities is being discussed and debated right now. But these questions depend fundamentally on the morally positive aspirational goals that we have developed. The current discussions of systemic racism and the Black Lives Matter movement are both, importantly, dependent on the stated (but incompletely lived up to) aspirational goals of American society.

The aspirational goals that we choose to teach each other matter. Well-designed aspirational goals can be powerful mechanisms to increase first-party self-control, can provide the opportunity to argue for second-party behavioral changes, and can shift third-party support. But what those aspirational goals are really do matter.

This was, I think, one of the points that the pastor I met on the airplane was talking about—that science would discover that the aspirational goals his religion provided him (help the needy, be forgiving, see the whole world as your community, and work to make it a better place for everyone) would be important for successful moral structures.

Aspirational goals are an important part of a moral structure. Well-designed aspirational goals can help increase cooperation within an inclusive community.

This is, in general, a good final summary of our story. Communities should be built so that everyone within that community will be happy participants, fulfilling their roles with gusto and satisfaction cooperating within the community without being selfish. The more people we can bring into that community, the better we will be. Thus, communities that reject others (that remove people who want to cooperate in the assurance game with them) are less moral than inclusive communities that bring more people into the shared community. To create that fair community, we need to assume that we do not know where we will end up within that community. Legal rules regarding what roles are available to us, what the punishments are for selfish behavior, and what support we have when we stumble should all be built ignorant of our initial position, our talents, and our luck.

Solving this question of community is playing the assurance game in the cooperate/cooperate corner, which is vastly more successful than living in the you're-on-your-own defect/defect corner. What's amazing is how well humans have worked to accomplish this. We call these solutions *morality*.

A NEW SCIENCE

This is only the beginning. In this book I have tried to provide a new framework for the science of morality. But within that framework, questions still remain.

For example, there are questions about how to handle information imbalances, where one party knows more about the situation than another, and questions about delayed information. *How would I have to change my behavior in an economic morality game if I wouldn't know you cheated until after a dozen rounds?*

There are experimental questions as to how these various moral structures interact with society as a community and with our individual decision-making systems as a whole. We looked at questions of police training paradigms and noted that they interact with decision-making systems that have real consequences. We talked about drug policy and the way that different views on addiction change the ease with which people can avoid and escape the addiction trap. But there are new questions that arise as technologies and societies change. For example: *How do the regulations of social media (or the lack of regulations) change our ability to remain in the assurance game? How do these various online forums interact with our decision-making systems?*

These are experimental sociological questions of morality. We can ask questions as to whether implementing a minimum wage or a universal basic income increases or decreases economic output, stability, and resilience across a community. Similarly, we can ask questions of how to deliver and fund social services. We saw that in some cases means testing was actually less efficient than just providing the resources to everyone. These are experimental questions that have actual answers we can measure. For example, we can ask: *What level of taxation is going to provide the largest infrastructure income to share? Which services are better run as a governmental community, and which are better run as private enterprise, given the different aspirational goals of government and private enterprise? Is it wise to put all social services under the rubric of an armed police force?* These experimental questions need to be addressed in the light of bringing ever-larger groups into our community of cooperators.

Furthermore, there are implementational questions as well. The data that masks prevent the transmission of COVID-19 is absolutely clear.[1] Wearing a mask protects both you and society. It is not only better for the community as a whole but better for you as an individual. "Wear a mask" is as obvious a morally positive statement as can be found. And yet, wearing a mask became a tribal marker in the US in 2020. Almost half of Americans refused to wear masks during the COVID-19 epidemic, and more than half a million Americans died of COVID-19 in 2020 and early 2021. That's a death rate of more than 1 in 500 and one of the highest national death tolls in the world.[2] *What went wrong with our moral structures here?*

This is the mark of a good science—it should open up new questions, even more than providing new answers. One such interesting question is how to handle the problem of climate change, where we have a common-pool resource, but the timescale is multigenerational. *How can we play an assurance game across generations? How can the interactional strategies developed for maintaining common-pool resources within an interacting community handle multigenerational games?* The question of climate change is very much a classic problem of the commons—the actions everyone takes affect it, and there is no easy way for us to escape the shared interactions of it. (We're not going to be able to be loners, avoiding the community here.) The complexity of climate change as a problem of the commons is that it is intergenerational.[3] The decisions and moral structures we implement now will affect us, yes, but they will have an even greater effect on generations to come.

We will need to do experiments, both at the individual level and at the society level, to determine the impact of these moral structures on behavior. We will need to conduct theoretical and mathematical analyses to determine how changes to the rules change the consequences of the games being played across generations and what moral structures we can build to maintain the integrity of our common-pool resources. Importantly, these consequences will depend on interactions with our decision-making systems and the specific ways they learn how and what to choose.

DECISION-MAKING SYSTEMS MATTER

Moral structures interact with our decision-making systems. Each of the decision systems learns differently. Each of the decision systems processes information differently. This means that how we train ourselves (how we provide learning) can change what individual systems learn to do. It also means that how we ask a question can change what answer we give.

We saw this interaction between decision systems and policy in questions such as the trolley problem, the Milgram experiment, and the ultimatum game. In the trolley problem, people were more willing to flip the switch to change the tracks than to push the person into the way of the train. Flipping the switch accessed deliberative decision systems, while pushing the person accessed Pavlovian decision systems. We saw this interaction between decision systems and policy in the way that empathy (a Pavlovian drive) can change our willingness to cooperate with another and in the way we can switch who we empathize with by changing our experiences with others growing up (and in our societies).

Understanding how training affects decision systems has important practical applications because these example problems (trolley problem, Milgram experiment) have real-world practical equivalents. How we train police officers has direct consequences on the interactions they have with individuals in our society. How we envisage our goals changes whether we are willing to apply torture or not.

Each of our decision systems can be taught to take actions that increase our likelihood of playing a cooperative role in an assurance game. Any social institutions we create will need to take these decision systems into account.

BISTABILITY AND RESILIENCE

And finally, one of the conclusions of this book is that the "moral arc bends towards justice." Groups made of people who work together, groups that work with other groups, and groups that support success for all members within the group are going to do better, on average, than groups that don't. But the actual quote from Martin Luther King Jr.'s famous speech is that the "arc of the moral universe is long, but it bends towards justice."[4] There is noise in the system. Sometimes empires collapse into anarchy. Sometimes democracies fall. But this does not change the science of morality. There are clear advantages to working together.

Our new science, however, has revealed a bistability to the assurance game. If you find yourself in a group of cooperators with a set of fair, helpful, moral structures and a set of aspirational goals of being cooperative members of that community, then you can be a cooperator within that community. It is better for the individual to cooperate within a cooperating community. (This is the difference between the assurance game and the prisoner's dilemma.) But if, on the other hand, you find yourself in a society of selfishists, where people are unwilling to cooperate with the community, then sharing your resources with them is a mistake. Remember, in the prisoner's dilemma you're a sucker if you cooperate while the other defects. Even in an assurance game, if the other player is going to defect, then cooperating is the wrong move.

Studies of repeated versions of the tax game find that predicted bistability. Some groups trend toward cooperation (ending up contributing a substantial amount of their individual resources to the central pot), while other groups trend toward holding on to their own money (ending up with almost no one contributing toward the central pot). These differences are (not surprisingly, given all we've seen so far) dependent on the moral structures included in the rules of the game.[5] In such situations, humans tend to find subgroups of cooperators with whom to bond into communities, which is a key explanation for the rise and collapse of empires and societies.[6]

However, our science has also revealed that we can create moral structures that interact with our decision systems to make us more likely to stay on the cooperation side of that bistability and keep us away from the defection side. We can create resilient societies of resilient individuals. We can create competition without conflict. We can create rules that can allow us to raise up the weak without tearing down the strong. We can, in the words of Paul Wellstone, all do better by all doing better. We can, in the words of Fred Rogers, learn to be neighbors.

I think the key is that we can change the rules of the game. Like Nick in that *Golden Balls* example in chapter 2, we can convince our opponent to be a partner and not an opponent, and we can convince them to split the pot with us even if they were originally planning to steal it. If we can change the rules of the game so we can be certain the other player is going to cooperate in an assurance game, then we can safely cooperate with them. We can work together as a group, and because the world is not zero-sum, we all win more as a group than we would as separate individuals.

In the end, the pastor I met on the plane was right. We are all part of a community, and there is a right and a wrong within that community that is based on helping others. He was right that, working together, if we create the right moral structures, we can find our way to share the increased gains from the cooperate-cooperate corner of the assurance game.

I have titled this book *Changing How We Choose: The New Science of Morality* because this is only the beginning. There is still a lot of work to be done. I hope that the ideas in this book engender a lot of discussion and maybe, just maybe, a step forward.

Acknowledgments

Every book (and every scientific discovery) is the product of a community. I could not have written this book alone. First and foremost, I want to thank my wife, Laura, my first reader, for her ability to push me on these ideas, for her willingness to read multiple drafts, and for her tolerance as I chased this obsession. Second, I want to thank my scientific colleagues, including my students (who are very much colleagues), who have helped me work out these ideas over the years. Their willingness to discuss and debate ideas, both in person and in print, has built a scientific community that I have been incredibly fortunate to be a part of. If I have put one brick in the edifice that we are building, it is because there has been a large edifice to build on. There are far too many people I have interacted with and talked to over the years to thank for including me in the community, but I do want to specifically thank C. Ford Runge, whose insights on the assurance game provided the final step into the start of this book. His willingness and generosity to reach across fields and find a common language with people outside his field are tremendous. I also want to thank the NeuroPRSMH group for providing a weekly discussion forum for addressing questions of psychiatry, decision-making, and a host of other topics closely related to the questions that I was grappling with in this book. I also want to thank Jan Dubinsky, Xiaosi Gu, Autumn Han, Steve Kelley, Angus MacDonald, C. Ford Runge, Francis Shen, Alik Widge, Iris Vilares, Sophia Vinogradov, and my parents, Joe and Ginny Redish, for reading parts (or all!) of the book. This book is better for their comments. I would also like to thank the anonymous reviewers who provided comments on the drafts of this book, which helped me improve it tremendously. Finally, I'd like to thank my editor Bob Prior, who gave me the right balance of room to move while still keeping the pressure on to make the book readable and accessible. I hope it is. All of the still-extant mstakes, of course, remain my own.

—David Redish, Minneapolis-Saint Paul, Minnesota, February 2022

Glossary

A note on game tables: In the tables below, Player A chooses a column, and Player B chooses a row, and the payout is in the corresponding entry. The return given to each player is noted by a right arrow (←). Thus, A←1, B←2 means that A gets one unit of reward, while B gets two units of reward, which the reader can interpret as \$1 and \$2, respectively.

asabiya: Social solidarity, the willingness to put one's group ahead of one's self.

assurance game: A two-player cooperate/defect game in which it is best to do what the other player did (cooperate if they cooperate and defect if they defect).

Assurance game		Player A	
		A chooses C	A chooses D
Player B	B chooses C	A ← 3; B ← 3	A ← 2; B ← 0
	B chooses D	A ← 0; B ← 2	A ← 1; B ← 1

chicken game: A two-player cooperate/defect game in which it is best to do the opposite of what the other player did (defect if they cooperate but cooperate if they defect).

Chicken game		Player A	
		A chooses C	A chooses D
Player B	B chooses C	A ← 2; B ← 2	A ← 3; B ← 1
	B chooses D	A ← 1; B ← 3	A ← 0; B ← 0

coalition defense: When one partner steps in to defend another partner against a third-party enemy. Note that coalition defense only requires "help my friend." Compare to *third-party punishment*.

collectivism: The idea that the goal of morality is what's best for the group despite what may be good for the individual.

compensation: Because we have multiple decision systems and how one asks a question changes which process drives behavior, a properly constructed social structure can allow an intact decision process to accomplish a task normally achieved through another (damaged) decision process.

deliberation: One of the major action-selection decision-making systems in human neurophysiology. Entails imagination of an explicit outcome that can be evaluated.

description: The philosophical concept in which one attempts to understand how the world works—*What are the consequences of actions and processes?* Compare to *prescription*.

descriptive theories: Identifying what components of an observation are repeatable.

dictator game: A two-player serial game in which Player A divides money between themself and Player B as Player A sees fit. Compare to the *ultimatum game*.

enemy at the gates: The problem of responding to an invading army or neighboring tribe.

enemy within: The problem of responding to a subgroup attempting to take control of a group.

excludable resources: A resource that one person can prevent another from using.

first-party self-control: When one person ensures that they themselves are going to behave morally, forgoing the opportunity to cheat.

framing: A psychological phenomenon whereby changes in the contextual situation and the constructs within which one envisages a question change the decisions made. It arises because of the way neural systems recognize patterns by pattern completion so external cues can shape what pattern the neural system settles on.

free rider: When an individual takes from the group without giving their fair share back.

game theory: The study of games in terms of states and actions and the maximization of outcome valuations.

gossip: Information shared among a group about an individual. See *reputation*.

harm reduction: The idea that the goal of morality is to minimize the harm that is done to individuals.

***Homo economicus*:** The idea that humans make choices to maximize value. Often used as a simple statement assuming maximization of resources; however, it is important to take into account preferences when identifying what value is, as well as the complexities of the decision-making systems, when applying the *Homo economicus* hypothesis.

instinctual decision system: see *Pavlovian decision system*.

intrinsic goals: Motivations that have been evolved to be endemic to the individuals of a species. By chasing those intrinsic goals, an individual is likely to succeed genetically without having to do the full (likely impossible) genetic calculation.

matching pennies game: A two-player game in which each player can choose one of two choices (heads/tails). Player A wins if both players pick the same choice. Player B wins if players pick different choices.

Matching pennies game		Player A	
		A chooses heads	A chooses tails
Player B	B chooses heads	$A \leftarrow 1; B \leftarrow 0$	$A \leftarrow 0; B \leftarrow 1$
	B chooses tails	$A \leftarrow 0; B \leftarrow 1$	$A \leftarrow 1; B \leftarrow 0$

mechanistic theories: An explanation of how something works as a structure/process of an interaction of smaller phenomena.

metanorms: Social norms that punishments and rewards will actually be meted out to ensure that primary social norms are followed.

moral injury: The idea that taking certain actions can create an internal conflict between one's moral view of oneself and the self one observes.

moral relativism: The idea that all moral systems are equally valid and that we cannot make statements about one being better than another.

naturalistic fallacy: The (incorrect) idea that whatever exists in the world is the best and right answer.

non-zero-sum game: A game in which it is possible for both players to win or both players to lose.

normative theories: An explanation of how well something works relative to an optimization goal.

nudge: A policy change that encourages one behavior over another but does not preclude either behavior.

operationalize: To define a construct or idea in terms of a measurable factor.

parochial altruism: Cooperation (altruism) within a group but xenophobia outside of it.

passive altruism: The observation that we don't take defection opportunities in normal circumstances that exist within our society.

Pavlovian decision-making: One of the major action-selection decision-making systems in human neurophysiology. Entails species-important behaviors that we already know how to do but can learn when to release.

prescription: The philosophical concept in which one attempts to define what one should do in certain situations. Compare to *description*.

priming: A psychological phenomenon whereby it becomes easier to recognize cues one is prepared for. It arises because of the way neural systems recognize patterns by pattern completion so being close to the pattern makes it easier to recognize.

prisoner's dilemma: A two-player cooperate/defect game in which defecting is always better than cooperating, but both cooperating is better than both defecting.

Prisoner's dilemma		Player A	
		A chooses C	A chooses D
Player B	B chooses C	$A \leftarrow 2; B \leftarrow 2$	$A \leftarrow 3; B \leftarrow 0$
	B chooses D	$A \leftarrow 0; B \leftarrow 3$	$A \leftarrow 1; B \leftarrow 1$

probe trial: An experimental manipulation that differentiates two potential explanations for a behavior.

problem of the commons: The question of how to manage shared resources, particularly those that are difficult to exclude people from using but which can be used up.

procedural decision-making: One of the major action-selection decision-making systems in human neurophysiology. Entails well-practiced action chains that can learn to be released (think sports).

public goods game: See *tax game.*

redemptive justice: Justice as a means of returning the offending individual back to society, which may require aspects of restorative justice and retributive justice, as well as apologies and contrition.

reputation: Information within a group about an individual, allowing interaction decisions to be made even upon the first encounter with an individual.

restorative justice: Justice as repairing the harm caused. Compare to *retributive justice.*

retributive justice: Justice as punishment. In retributive justice, the goal is to titrate the punishment to the crime. Compare to *restorative justice.*

rivalrous resources: A resource that only one person can use at a time.

second-party dyadic punishment: When one person punishes another for being unfair to them.

signaling: Providing social information about one's intentions. (Importantly, signals are not necessarily truthful.)

sociobiology fallacy: The (incorrect) assumption that current cultural observations are due to completed optimization processes and thus are the optimal choices available.

tax game: A multiplayer game in which players put money into a central pot, which then increases and pays out equally to all players regardless of how much (if anything) they contributed to the central pot. Also known as the *public goods game*.

third-party punishment: When a member of a team punishes another member of the same team for cheating a third member of the team.

tit for tat: A strategy in a repeated two-player game that cooperates on the first interaction but then does whatever the other player did in the previous round.

tragedy of the commons: See *problem of the commons*.

transactional costs: Costs paid to complete an interaction that are lost once the interaction is complete.

tribal marker: A signal that one is part of a specific group or subgroup. Tribal markers can be innocuous (such as knowledge of a secret hand signal) or dangerous (such as an unwillingness to get vaccinated for a disease).

trustee game: A two-player game in which Player A can pass some portion of their money to Player B. The money increases (triples) on the way to Player B, who can then return a portion of the increased money to Player A.

ultimatum game: A two-player game in which Player A proposes a division of money between themself and Player B. Player B can then "take it or leave it." If Player B accepts the deal, the players go home with the split that Player A proposed. If Player B rejects the deal, no one gets anything. Compare to the *dictator game*.

utilitarianism: The philosophy that the moral goal is to maximize the total happiness over the most people.

zero-sum game: A game in which if one side wins, the other loses.

Notes

CHAPTER 1

1. **Moral decisions:** Damasio, 1994, 2003; Moll et al., 2001; Greene et al., 2001; Greene, 2013; Sanfey et al., 2003; Sanfey, 2007; Haidt, 2012; Crockett, 2017.

2. **Humans handle moral questions differently:** Goodall, 1986; Jensen et al., 2007; Riedl et al., 2012; Proctor et al., 2013; Henrich and Silk, 2013; Tomasello, 2016.

3. **Philosophies in neuroscience integrations:** Churchland, 2011, 2019; Greene, 2013.

4. **Altruism as genetic success:** Hamilton, 1964; Alexander, 1974; Dawkins, 1976.

5. **Ants and bees:** Wilson, 1980; Queller and Strassmann, 1998.

6. **Neptune and Mercury:** Levenson, 2015.

7. **Apollo 13:** J. Lovell and J. Kluger (1994), *Lost Moon: The Perilous Voyage of Apollo 13*, Houghton Mifflin; and the movie *Apollo 13*, directed by Ron Howard (Universal Pictures, 1995). For a wonderful fictional account of how changing goals changes the application of science, see *The Martian* (the original book by Andrew Weir [2011] and the movie directed by Ridley Scott [2015]), in which Mark Watney says, after being stranded alone on Mars, that he is going to have to "science the shit out of this" in order to survive.

8. **Science is slow:** Contopoulos-Ioannidis et al., 2008; Redish et al., 2018.

9. **We all do better when we all do better:** P. Wellstone (1998), Speech before the 99th National Convention of the Veterans of Foreign Wars.

10. **Monetary tools can backfire:** Gneezy and Rustichini, 2000; Binmore, 2005; Bowles, 2016.

    Chapter sources: This chapter draws from a number of literatures, including the mathematics of economic games (Taylor, 1976; Axelrod, 1984; Sugden, 1986; Kollock, 1998; Skyrms, 2004; Binmore, 2005; Curry, 2016), the mathematics of communities (Runge, 1981, 1984a, 1984b; Ostrom, 1990; Ellickson, 1994; Kollock, 1998; Sober and Wilson, 1998; Fehr and Fischbacher, 2004; Wilson, 2010, 2015; Bowles and Gintis, 2011), and anthropology and evolutionary biology (Goodall and van Lawick, 1971; Goodall, 1986; Strum, 1987; Cheney and Seyfarth, 1990; de Waal, 1989; Boyd et al., 2003; Boehm, 2012; Tomasello, 2016; Hare and Woods, 2020), including studies of religion (Boyer, 2001; Atran, 2002; Wilson, 2015) and history (Turchin, 2003, 2006, 2016; Wright, 1999; McCullough, 2020) and conceptualizations of social construction (Schelling, 1960; Searle, 1995; Enfield and Levinson, 2006; Pinker, 2011; Young, 2015; Henrich, 2016). It also draws from the neuroscience and psychology of decision-making (Redish, 2013b), particularly in the context of morality (Damasio, 1994, 2003; Rai and Fiske, 2011; Haidt, 2012; Greene, 2013; Geşiarz and Crockett, 2015; Crockett, 2017; Churchland, 2019). Finally, for discussions of moral philosophy and legal theory, see Aristotle (350 BCE/1925); Augustine of Hippo (Saint Augustine; 427/1972, 398/1961); Hobbes (1651); Locke (1690); Kant (1781/1787); Hume (1777/1912); Rawls (1971, 1985); Scanlon (1998); Shermer (2004, 2015); Hoffman (2014). The "What Is a Science?" section builds on conceptualizations of science laid out by Firestein (2012, 2015), Ben-Ari (2011), and Laudan (1978), as well as a recent manuscript I have been working on with a number of my colleagues (available in arXiv; Levenstein et al., 2020).

CHAPTER 2

1. **Operation Fortitude:** J. Levine (2011), *Operation Fortitude: The Story of the Spies and the Spy Operation That Saved D-Day*, Lyons Press; R. Beyer and E. Sayles (2015), *The Ghost Army of World War II*, Princeton Architectural Press.

2. **Defect more on television:** Burton-Chellew and West, 2012; Capraro et al., 2014; Heuer and Orland, 2019.

3. **Ibrahim and Nick:** *RadioLab* (2014), *The Golden Rule*, WNYC.

4. **Ibrahim interview:** Ibid.

    Chapter sources: The ideas in this chapter build from game theory (von Neumann and Morganstern, 1944; Camerer, 2003; Osborne, 2004) and from the "Mesoeconomics" section of the bibliography. The prisoner's dilemma was introduced by Runciman and Sen (1965) and Rapoport and Chammah (1965). See Kollock (1998) and Poundstone (1992) for good reviews.

CHAPTER 3

1. **Tragedy of the commons:** Lloyd, 1832; Hardin, 1968.

2. **Fishery limits:** Gissurarson, 2000; Burger et al., 2001; Hamilton et al., 2004.

3. **Enclosure:** Allen, 1982.

4. **Taxes:** See Kelton (2020) for a discussion of how taxes differ between governments that issue their own currency vs. governments that don't.

5. **Tax rates:** Tax Policy Center, www.taxpolicycenter.org.

6. **Altruism as self-interest:** Hamilton, 1964; Alexander, 1974; Dawkins, 1976; Wright, 1995; Cronk and Leech, 2013; Joyce, 2007. The quote is from Ghiselin, 1974.

7. **Seasonal cooperation:** Sober and Wilson, 1998. *General property of interactions across levels:* Wilson, 2015.

8. **Conditional cooperation:** Marwell and Ames, 1981; Fehr and Fischbacher, 2004.

9. **Punishment increases cooperation:** Yamagishi, 1986; Ostrom et al., 1992; Fehr and Fischbacher, 2004; Rockenbach and Milinski, 2006; Bowles and Gintis, 2011.

10. **Group definitions matter:** Ostrom et al., 1994; Turchin, 2003; Bowles and Gintis, 2011; Wilson, 2015.

11. **Reputation and communication matter:** Roth and Schoumaker, 1983; Axelrod and Dion, 1988; Emler, 1994; Ostrom et al., 1992; Fehr and Fischbacher, 2004; Rockenbach and Milinski, 2006.

Chapter sources: The stag hunt example goes back to Rousseau (1762). The assurance game has been explored extensively within the mesoeconomics literature (see reviews by Kollock, 1998; Sugden, 1986; Skyrms, 2004). Altruism as genetic self-interest goes back to Hamilton (1964). See Wright (1995) for a popular science review. For discussion of the commons and its issues, see work by Runge (Runge, 1981, 1984a; Runge and Defrancesco, 2006) and Ostrom (Ostrom, 1990; Ostrom et al., 1994; Burger et al., 2001). The observation that a repeated prisoner's dilemma is an assurance game goes back to Taylor (1976) and Axelrod (1984). The tax game has been extensively studied experimentally, usually under the rubric of the "public goods" game (Dawes, 1980; Marwell and Ames, 1981; Ostrom et al., 1992; Fehr and Fischbacher, 2004; Smith, 2008; McElreath and Boyd, 2007; Bowles and Gintis, 2011; Henrich and Muthukrishna, 2021; van Dijk and De Dreu, 2021).

Notes on economic games: Technically, three very different two-player games can be explored here: the prisoner's dilemma, the assurance game, and the *chicken game*. We've seen real-world examples of the prisoner's dilemma and assurance games. The chicken game is played, for example, by driving two cars toward each other at high speed, and the first one to swerve loses. If one player swerves (cooperates), then the other can keep driving, gaining "face" for being tougher. If neither player swerves, then the cars hit head on and both players die. Similarly, there are variations in the multiplayer group games to be explored, the differences of which depend on how quickly the return grows as the number of players putting money into the pot changes. For example, optimal strategies change whether the growth function is concave (accelerates early quickly, making early contributions more important), is convex (accelerates late, making late contributions more important), is a simple linear function (as we explore here), or is a step function (like a Kickstarter campaign, which is only triggered after enough people have contributed or enough money has been promised). Kollock (1998) has an excellent review of these issues, including how the different two-player payout matrices change optimal decisions and the consequences of these different growth functions. While fascinating, these issues are beyond the scope of the argument made in this book—that science has a handle on the question of morality and that lots of good science remains to be done here.

CHAPTER 4

1. **Evolution is about the distributions of traits**: Darwin, 1859; Gould, 2002; Mayr, 2008; Wiener, 1995.

2. **Primates**: Goodall and van Lawick, 1971; Goodall, 1986; de Waal, 1982, 1989; Cheney and Seyfarth, 1990; Sapolsky, 2001; Strum, 1987.

3. **Bats**: Carter and Wilkinson, 2013; Carter et al., 2020.

4. **Changing how one got the money affects the ultimatum game**: Bland et al., 2017.

5. **Playing against a person**: Blount, 1995.

6. **Changing beliefs**: *Share less:* Rand, 1957; Marwell and Ames, 1981; Fox, 2009. *Share more:* Wilson, 2010; Bregman, 2020. For discussion and reviews, see Marwell and Ames, 1981; Stout, 2008, 2011; Fox, 2009; Wilson, 2010; Bregman, 2020.

7. **Changing thinking about sharing**: Rand, 2016; Bland et al., 2017.

8. **Cultural dependence**: Oosterbeek et al., 2004; Greene, 2013; Henrich, 2016; Bowles and Gintis, 2011; Bowles, 2016.

9. **Changing community**: Rachlin and Jones, 2008b; Locey et al., 2013; Marsh, 2017.

10. **Neurophysiological and pharmacological manipulations in the ultimatum game**: Sanfey et al., 2003; Sanfey, 2007; Fehr and Camerer, 2007; Krajbich et al., 2009; van den Bos and Güroğlu, 2009; MacDonald and MacDonald, 2010; Müller-Leinß et al., 2018; Terris et al., 2018; Speitel et al., 2019.

11. **Increasing trust**: Berg et al., 1995; King-Casas et al., 2008.

    Chapter sources: A repeated game is an assurance game (Taylor, 1976; Axelrod, 1984; Sugden, 1986). The ultimatum game was introduced by Güth et al. (1982), the dictator game by Forsythe et al. (1994), and the trustee game by Kreps (1990; see also Berg et al., 1995). See Alos-Ferrer and Farolfi (2019) for reviews.

CHAPTER 5

1. **Bees**: Hamilton, 1964; Hölldobler and Wilson, 1990; Queller and Strassmann, 1998.

2. **Home**: The original quote is from Robert Frost's poem "The Death of the Hired Man" (1914), but I have to admit I had forgotten that and pulled it from Lois McMaster Bujold (1990), *The Vor Game*, Baen Books.

3. **Chickens**: Gottier, 1968.

4. **Soft power**: Strum, 1987; Byrne and Whiten, 1989; de Waal, 1982, 1989; Sapolsky, 2001.

5. **Cannae**: O'Connell, 2010.

6. *Watchmen*: In the graphic novel (A. Moore and D. Gibbons [1987], DC Comics), Adrian Veidt simulates an alien invasion killing millions through psychic horror, while in the movie (Z. Snyder [2009], Warner Bros./Paramount), he simulates destruction attributed to

Dr. Manhattan. Whether Dr. Manhattan is technically a human with superpowers or whether his superpowers have effectively made him an alien, I will leave for the inevitable discussion on the internet. (The question of his humanity is an important trope in both the novel and the movie.) In any case the existence of a common enemy creates world peace in both the novel and movie.

7. **Primates:** Goodall, 1986; Strum, 1987; de Waal, 1982, 1989; de Waal and Lanting, 1997; Cheney and Seyfarth, 1990; Sapolsky, 2001.

8. **Limited evidence for third-party punishment in nonhuman primates:** Riedl et al., 2012; von Rohr et al., 2012; Suchak et al., 2016; Tomasello, 2016.

9. **Passion and Pom:** Goodall, 1990.

10. **Robbing banks:** Reilly et al., 2012.

11. **Parole:** US Department of Justice Parole Commission (2020), *Parole System.*

12. **Conversation/tax game:** Fehr and Fischbacher, 2004; Rockenbach and Milinski, 2006; Milinski, 2016; Hare and Woods, 2020.

13. **Gossip:** Emler, 1994; Dunbar, 1996; Foster, 2004; Boehm, 2012.

14. **Gossip and hypocrisy:** Emler, 1994; Boehm, 2012; Jordan et al., 2017.

15. **Experiments on cheating:** Ariely, 2008; Bucciol and Piovesan, 2011; Fischbacher and Föllmi-Heusi, 2013.

16. **Coffee service experiment:** Bateson et al., 2006.

17. **Atheists are moral:** Anderson and Mellor, 2009; Grossman and Parrett, 2011.

Chapter sources: Asabiya was introduced to the modern literature by Turchin (2003, 2006, 2016). *Gene as atom of evolution:* Hamilton, 1964; Alexander, 1974; Dawkins, 1976; see Wright (1995) for a popular review. *Cultural transmission:* de Waal, 2001, 2009; Bowles and Gintis, 2011; Henrich, 2016. See Hoffman (2014) for a discussion of crime and punishment and Haidt (2012) for cross-cultural comparisons of what humans define as crimes. See Boehm (2012) for the anthropological literature. For studies of third-party punishment in the tax game, see the mesoeconomics literature (Yamagishi, 1986; Ostrom, 1990; Ostrom et al., 1992; Fehr and Fischbacher, 2004; Bowles and Gintis, 2011; van Dijk and De Dreu, 2021). Katchadourian (2011) explores the constructs of guilt, shame, and regret. Reputation and gossip are discussed in both the mesoeconomics and anthropological literature: Roth and Schoumaker, 1983; Merry, 1984; Rockenbach and Milinski, 2006; Boehm, 2012; Milinski, 2016. *Moral praise and moral blame:* see Anderson et al. (2020) for a recent review. *Religion as imagined surveillance:* see Boyer (2001) and Wilson (2010) but also Atran (2002) for an alternate view.

CHAPTER 6

1. **Studies of actual communities:** *Fisheries:* Ostrom, 1990; Burger et al., 2001. *Maine lobstermen:* Acheson, 1988. *Nineteenth-century whalers:* Ellickson, 1994. *Indigenous hunter-gatherer*

bands: Boehm, 2012. *Northern California ranchers:* Ellickson, 1994. See also Bowles and Gintis (2011); Diamond (2013); Wilson (2015).

2. **Monetary compensation/culture:** Boehm, 2012. The Norse and Icelandic sagas show several examples of monetary compensation for crimes. See, for example, *Egil's Saga* (Sturluson, 1230), *The Laxdæla Saga* (unknown author, 1245), and *Njal's Saga* (unknown author, 1279) for three examples.

3. **Haifa day care center:** Gneezy and Rustichini, 2000; Bowles and Gintis, 2011; Bowles, 2016.

4. **Revenge:** See the Norse and Icelandic sagas in note 2 for classic examples of these revenge cycles turning into feuds.

5. **Ability to punish increases cooperation:** Yamagishi, 1986; Ostrom, 1990; Fehr and Fischbacher, 2004; Bowles and Gintis, 2011; van Dijk and De Dreu, 2021.

6. **Third-party vs. second-party punishment:** Traulsen et al., 2012.

7. **Cheating in the hunt:** Boehm, 2012.

8. **Less willing to cooperate outside of one's group:** Wilson, 2010; Boehm, 2012; Diamond, 2013; Greene, 2013; Hare and Woods, 2020.

9. **Nonviolence quote:** Quoted in Chenowith and Stephan (2011). See also Malcolm X and A. Haley, *The Autobiography of Malcolm X*, Grove Press.

10. **Revolutions:** Chenowith and Stephan, 2011.

11. **Convinced by aspirational goals:** *Lincoln:* Foner, 2010. *Johnson:* Caro, 2013; Lewis, 2012.

Chapter sources: *Sociological studies of lobstermen in Maine:* Acheson 1988. *Ranchers in Northern California:* Ellickson 1994. Additional ideas in this chapter build on work by Boehm (2012), de Waal (1982, 1989), and Tomasello (2016). *Economics of social control:* Ostrom, 1990; Bowles and Gintis, 2011; Wilson, 2015; Henrich and Muthukrishna, 2021; van Dijk and De Dreu, 2021. *Changes in social constructions with group sizes:* Wright, 1999; Pinker, 2011; Diamond, 2013; Turchin, 2016; Hare and Woods, 2020; McCullough, 2020. *Success of violent and nonviolent movements:* Chenowith and Stephan, 2011. Nonpositive norms can be seen in descriptive discussions of community codes by Fiske and Rai (2015) and Haidt (2012). See Wilkerson (2020) for an insightful discussion of the consequences of these codes on what she calls "the dominant caste." See also McGhee (2021).

Note on the interaction between third-party interference and aspirational goals: One of my favorite literary examples of the way that aspirational norms, nonviolence, and third-party punishment interact is from the TV show *Deep Space Nine* episode "The House of Quark," in which the Ferengi Quark finds himself in an honor duel with the Klingon D'Gohr in front of the Klingon high council. Quark knows that he has no chance of surviving this fight. Quark is physically half the size of D'Gohr and has no military training, while D'Gohr has been practicing for these duels his whole life. But in the show, it had long been established that in Klingon culture killing an opponent in battle is honorable, but slaughtering civilians is not. So Quark throws down his *Bat'leth* (the special Klingon sword they are using to fight with) and kneels down to be executed. D'Gohr then steps back to deliver what would be a killing blow, but the council leader steps in to stop him. After stopping the fight, the Klingons ostracize D'Gohr, stripping him of his rights, lands, and honor, dramatically shown as they turn their backs to him as guards come to throw him out of the council chamber. (Technically, they give him a "discommendation," which had been established in an earlier episode as a removal of rights from the community.) Quark, through this dramatic scene, is able to prove that D'Gohr is unwilling to play by the norms of the community and actually wins the battle against a significantly stronger enemy.

There is much to unpack in this scene between Quark and D'Gohr. Quark uses nonviolence (throwing down his sword, laying down his life) to win the battle. In doing so, Quark has changed the game. D'Gohr is fighting a dyadic duel, where he has to defeat his opponent. Quark is playing a triadic game, depending on D'Gohr's inability to recognize the importance of community. Notice how the success of Quark's nonviolence depends on the community he is playing to. The council is shocked and angry at D'Gohr's willingness to kill an unarmed opponent. Such behavior is not acceptable within the Klingon community. Quark uses the moral force of aspirational norms that all Klingons should fight with honor and that killing an unarmed opponent half your size is not an honorable duel. It is important to note that Quark's dependence on (use of) the physical strength of the other Klingons in the community (particularly the high council they are dueling in front of, which outranks D'Gohr) does not diminish the fact that in the end Quark wins and D'Gohr loses.

CHAPTER 7

1. *The Mind within the Brain*: Redish, 2013b.

2. **Nudge:** Thaler and Sunstein, 2008.

3. **Simulations in our head:** Johnson and Redish, 2007; Buckner and Carroll, 2007; Gilbert and Wilson, 2007; Schacter et al., 2007; Doll et al., 2015; see Redish (2016) for a review.

4. **Pattern completion:** Hebb, 1949/2002; Hertz et al., 1991.

5. **Deep learning:** LeCun et al., 2015; Goodfellow et al., 2016.

6. **Expertise:** Ericsson et al., 1993; Klein, 1999; Ericsson, 2006.

7. **Donating a kidney to a stranger:** Marsh, 2017.

8. **Episodic memory as a decision process:** Stewart et al., 2006; Shadlen and Shohamy, 2016.

9. **Situation categorization as a decision process:** Gold and Shadlen, 2001; Striedter, 2005; Redish et al., 2007.

10. **Decisions based on rules:** Johnson-Laird, 2010; Kahneman, 2011; Haynos et al., 2022. A rule-based decision system will have moral and sociopolitical decision-making consequences based on what rules one decides to follow (March and Olsen, 2011). See Koechlin et al. (2003) for interesting data on increasing abstraction as one moves more frontal in the human neocortex.

11. **Dionysus and Apollo:** Dionysus was the god of wine and is often portrayed as a party animal frat boy, while Apollo was portrayed as the cold intellectual. However, one of the most interesting things about Indo-European mythology, including Greek mythology, is how human and multidimensional the gods are. There were definite myths in which Dionysus was cold and calculating (as in the Midas story) and times when Apollo was full of wild Pavlovian lust (as in the Daphne story). Nevertheless, Plato and others used the concepts of hot and cold and associated them with Dionysus and Apollo, respectively.

12. **Animal spirits:** Keynes, 1936; Akerlof and Kranton, 2010; Coates, 2012.

13. **Priming:** Desimone, 1996; Wiggs and Martin, 1998; Bargh and Chartrand, 2014.

14. **Multiple decision systems:** I've reviewed the important points in this chapter, but interested readers can find more details in Redish (2013b).

15. **Fast is not always better:** See the discussion of expertise in Klein (1999) and Groopman (2007) and the discussion of when Pavlovian and procedural systems are more appropriate than deliberative in Redish (2013b).

16. **Horse-and-rider theories:** Haidt, 2006; Eagleman, 2011; Gazzaniga, 2012, 2019; Kurzban, 2010.

17. **Dissociation:** Classen et al., 1993; Feeny et al., 2000.

    Chapter sources: The main concepts in this chapter derive from the neuroscience and psychology section of the bibliography and a long history of recognizing that the decision-making process in humans is not unitary. Those main ideas are reviewed in depth in Redish (2013b).

    Note on multiple decision systems: The fact that we are not unitary decision-makers but rather a federation of action-selection algorithms implies that there must be a process to determine which system controls actions when. We currently don't know what that process is. A number of theories have been proposed, but they are not critical to the questions we are addressing here.

CHAPTER 8

1. **Importance of equality before birth for moral decisions:** Rawls, 1971, 1985.

2. **Milgram replication:** Burger, 2009; Burger et al., 2011. See also Worchel and Brehm (1970) for psychological theories of attitude and Miller (2009) for a discussion of the differences between the Burger and Milgram studies.

3. **Batson study:** Batson et al., 1981.

4. **Milgram and gender:** Burger, 2009. A similar (but not identical) Batson et al. (1981) study used solely women undergraduates. Of course, American societal gender norms and roles were very different in 1961, 1981, and 2009 and are even more different today in 2021.

5. **Americans have positive views of science:** Pew Research (2020), *Key Findings about Americans' Confidence in Science and Their Views on Scientists' Role in Society.*

6. **A hierarchy of relationships:** Jones and Rachlin, 2006; Rachlin and Jones, 2008a; Locey et al., 2013; Marsh, 2017.

7. **Trolley problem in a second language:** Costa et al., 2014.

8. **fMRI:** Huettel et al. (2004) is a good textbook with technical details. Briefly, fMRI measures changes in the oxygen levels in blood. The active parts of the brain require more blood flow but don't require more oxygen from the blood, so the increased blood flow has a higher ratio of oxygenated to deoxygenated hemoglobin, which is detectable because having oxygen attached to a hemoglobin molecule changes its response to magnetic fields.

9. **Trolley problem and neural circuits:** Moll et al., 2001; Greene et al., 2001.

10. **Distinctions between neural circuits:** *Food decisions:* Hare et al., 2009; Rangel and Hare, 2010. *Marshmallow task:* Mischel, 2014; McClure et al., 2004; Bechara, 2005. *Ultimatum games:* Sanfey et al., 2003; Singer, 2008; Cooper et al., 2010; Greene, 2013.

11. **Unwillingness to kill:** Laidley, 1865; Marshall, 1947; Grossman, 1995; Jordan, 2002; Bregman, 2017. Although the specifics of the original study by S. L. A. Marshall from World War II have been called into question, subsequent studies have supported the conclusions.

12. **Killing changes one psychologically:** Shay, 1994, 2002; Grossman, 1995; Fontana et al., 1992; Maguen et al., 2011; Bateson, 2015.

13. **Killing at a distance:** Grossman, 1995; Campo, 2015.

14. **PTSD:** Chappelle et al., 2014. See also US Department of Defense Health Agency (2017), *Mental Health Disorder Prevalence among Active Duty Service Members in the Military Health System, Fiscal Years 2005–2016.*

15. **The self and noncognitive systems:** *Press secretary:* Kurzban, 2010. *Superficial shell:* Gazzaniga, 2012, 2019; Haidt, 2006. *Zombie processes:* Eagleman, 2011.

16. **Bystanders help:** Fischer et al., 2011; Philpot et al., 2019.

17. **Kitty Genovese:** See Bregman, 2020.

18. **Expertise:** Klein, 1999; Groopman, 2007. See Redish (2013b) for a detailed discussion of the computational, algorithmic, and neural differences between procedural and Pavlovian decision systems.

19. **Policing:** US President's Task Force on 21st Century Policing (2015), *Final Report of the President's Task Force on 21st Century Policing.* See also Stoughton (2016); Delehanty et al. (2017); Mummolo (2018); McElrath and Tuberville (2020); Nordell (2021); Alexander (2010); and DuVernay (2016) for extensive discussions of these topics.

20. **Militarized training of Officer Yanez:** Grossman (1995) cites clear data that the natural tendency for humans is not to kill, particularly at close range. Grossman then argues that soldiers need to kill their enemies, which is, of course, not the actual ultimate goal of military strategy. (The actual goal of military strategy is to win the war.) He then argues that police are soldiers in a war (which belies the roles actually identified for police as playing a part in helping society be a community) and begins touring the country, providing military training

to police officers to learn to kill more easily. The scientific logic of morality that I have been developing in this book identifies just how problematic Grossman's conclusions are from a moral perspective. Yanez had recently completed one of these training seminars. See B. Schatz (March/April 2017), Are you prepared to kill somebody? *Mother Jones*, https://www.motherjones.com/politics/2017/02/dave-grossman-training-police-militarization. For a scientific study of the role of Pavlovian decision systems in police shootings in general, see Hashemi et al. (2019).

21. **PTSD and suicidality**: Fontana et al., 1992; Maguen et al., 2011; Bateson, 2015.

22. **Neural circuits of third-party punishment**: Seymour et al., 2007; Singer, 2008.

    Chapter sources: Milgram's experiments are described in depth in Milgram (1974). *Additional details and interpretations of Milgram:* Morelli, 1983; Haslam and Reicher, 2007; Miller, 2009; Reicher and Haslam, 2011; Russell, 2011; Bregman, 2020. *Trolley problem:* Foot, 1967; Greene, 2013; Bauman et al., 2014; Bruers and Braeckman, 2014; Edmonds, 2015.

CHAPTER 9

1. **Lissek experiments**: Lissek et al., 2014; van Meurs et al., 2014.

2. **Sugar**: Sapolsky, 1998.

3. **Long-distance running**: Bramble and Lieberman, 2004; Liebenberg, 2012.

4. **WEIRD**: There is a backlash against studies that involve people from Western, educated, industrialized, rich, and democratic societies (note the acronym WEIRD) in discussions of both psychology (Henrich et al., 2010; Henrich, 2016) and morality (Haidt, 2012). While I agree that one cannot identify universals from any limited sociological background, I take exception to the idea that it is invalid to study these societies. These are human societies. Billions of people live in them. As such, they are as valid as any other to understand the human condition. That being said, I would expect the new scientific definition of morality that I am laying out in this book to apply to all societies, universally, WEIRD and not. We are all trying to find moral codes that will allow us to reach the cooperation corner of the assurance game and reap the non-zero-sum benefits therein.

5. **Reacting to poor deals in the ultimatum game with emotion**: Sanfey et al., 2003; Sanfey, 2007.

6. **Playing fairer with humans than computers**: Blount, 1995; Fehr and Fischbacher, 2004; Bolton et al., 2016.

7. **Fairness in nonhuman primates**: Leimgruber et al., 2016; de Waal, 2014; Tomasello, 2016.

8. **Ultimatum games and neural circuits**: Sanfey et al., 2003; Sanfey, 2007; Greene, 2013; King-Casas et al., 2008.

9. **Changing decisions in ultimatum games**: Rand, 2016; Bland et al., 2017.

10. **Retributive justice**: Hoffman, 2014; Acheson, 1988; Ellickson, 1994; Boehm, 2012.

11. **Expertise develops as procedural decisions and perceptual pattern recognition**: Klein, 1999; Redish, 2013b.

12. **Drug treatment better than jail**: Manski et al., 2001; Leukefeld et al., 2011; Meyer and Mirin, 1979; Petry, 2012; Hart, 2013.

13. **Solitary confinement**: Hagan et al., 2018; Coppola, 2019.

14. **Loneliness increases mortality**: Cacioppo et al., 2014; Xia and Li, 2018; Beller and Wagner, 2018.

15. **Familiarity can decrease boundaries**: Allport, 1954; Nordell, 2017, 2021; Bregman, 2020; McCullough, 2020.

16. **Status effects on reproduction differ between men and women**: Hrdy, 1999.

17. **Complexities of primate social structure**: Kummer, 1971; Goodall, 1986; de Waal, 1982, 1989; Fossey, 1983; Strum, 1987; Ghiglieri, 1988; Cheney and Seyfarth, 1990; Hrdy, 1999; Sapolsky, 2001; Boesch and Boesch-Achermann, 2012.

18. **R.E.S.P.E.C.T**: A. Franklin (1967), "Respect," Atlantic Records.

19. **Tool use**: *Caddisfly larva:* Dawkins, 1976. *Primates:* Kawai, 1965; Kummer, 1971; Goodall, 1986; Boesch and Boesch-Achermann, 2012. *Humans:* Diamond, 1997, 2013; de Waal, 2001; Enfield and Levinson, 2006; Henrich, 2016.

20. **Cultural transmission**: de Waal, 2001; Enfield and Levinson, 2006; Henrich, 2016.

21. **Extra steps**: *Humans:* Lyons et al., 2007; Nielsen et al., 2008. *Other primates:* Call et al., 2005; Tomasello, 2016.

22. **Humans imitate each other**: DePaulo, 1992; Chartrand and Bargh, 1999.

23. **Reporting what the group says rather than reality**: Asch, 1948, 1952; Festinger, 1957; Janis, 1982; Hodges and Geyer, 2006.

24. **Pop-up shooter training**: Grossman, 1995.

25. **Most soldiers don't shoot**: Marshall, 1947; Grossman, 1995.

Chapter sources: The concept of intrinsic goals comes from work by Edelman (1992), Heckhausen (2000), and Uchibe and Doya (2008). I also addressed intrinsic goals in *The Mind within the Brain*, where I referred to them as *intrinsic goal functions* (Redish, 2013b). Many of the specific intrinsic goals are described in the "Ethology, Anthropology, and Evolutionary Biology" section of the bibliography, particularly the work on parochial altruism (Turchin, 2003, 2016; Boehm, 2012; Haidt, 2012; Diamond, 2013; Tomasello, 2016; Hare and Woods, 2020). See also punishment as an intrinsic goal (Haidt, 2012; Hoffman, 2014), fairness (Seymour et al., 2007; Boehm, 2012; Haidt, 2012; Tomasello, 2016), status (Goodall, 1986; de Waal, 1982; Sapolsky, 2001, 2017; Diamond, 2006, 2013; Boehm, 2012; Wilkerson, 2020), within-group controls to prevent defection (Wilson, 2010, 2015; Boehm, 2012; Tomasello, 2016), and the importance of a support structure and the avoidance of loneliness (Bowlby, 1969; Harlow, 1986; Cacioppo et al., 2014, 2015; Xia and Li, 2018; Beller and Wagner, 2018; Hagan et al., 2018; Coppola, 2019; Tomova et al., 2020).

CHAPTER 10

1. **People solve the assurance game with empathy**: Batson, 2010, 2014; Moll et al., 2006; Harbaugh et al., 2007; Hare et al., 2010; Tusche et al., 2016.

2. **Imagination uses the same neural systems as perception**: Kosslyn, 1994; O'Craven and Kanwisher, 2000; Rizzolatti and Craighero, 2004; Pearson et al., 2015.

3. **Empathy as imagination of emotion**: Singer, 2008; Bastiaansen et al., 2009; Marsh, 2017.

4. **Being rejected socially hurts**: Eisenberger et al., 2003; Kross et al., 2011.

5. **Betraying someone you empathize with hurts**: Aimone et al., 2014.

6. **Experience with diversity increases tolerance**: Allport, 1954; Nordell, 2017, 2021; Wilkerson, 2020; McCullough, 2020.

7. **Neurophysiology of trust**: Sanfey et al., 2003; King-Casas et al., 2005; Singer, 2008; Koscik and Tranel, 2011; Fareri et al., 2012; Chang et al., 2013; Aimone et al., 2014; Marsh, 2017; Elmontaite et al., 2019. *Amygdala in Pavlovian actions:* Maren and Quirk, 2004; LeDoux, 1996; Janak and Tye, 2015. *Amygdala in social situations:* Sanfey et al., 2003; Gothard et al., 2017; Marsh, 2017. *Ventromedial and orbitofrontal cortices:* Damasio, 1994; LeDoux, 1996; Barrett, 2017. *Anterior insula*: Singer, 2008; Sanfey et al., 2003; Aimone et al., 2014.

8. **Psychiatric disorders and trust games**: King-Casas et al., 2005; King-Casas and Chiu, 2012; Kishida et al., 2010; Marsh, 2017.

9. **Super-altruists**: Marsh, 2017.

10. **Oxytocin**: In mammals (including humans), there are two such chemicals, oxytocin and arginine vasopressin. They have similar (but not identical) chemical structures and subtly different effects (MacDonald and MacDonald, 2010; Neto et al., 2020).

11. **Oxytocin and pets**: Feldman, 2012; Rilling, 2013.

12. **Oxytocin and voles**: Young et al., 1998; Insel and Young, 2001.

13. **Oxytocin nasal delivery**: Because odor receptors are neurons that must interact with chemicals in the outside world (that's what it means to smell), a chemical provided to the nose can move directly to the brain. Therefore, oxytocin can be provided nasally to determine what effect it has on human brain function (Neumann et al., 2013; Evans et al., 2014).

14. **Oxytocin and ultimatum/dictator games**: Zak et al., 2007; Neto et al., 2020.

15. **Oxytocin and trustee games**: Kosfeld et al., 2005; MacDonald and MacDonald, 2010.

16. **Oxytocin receptor distributions in humans**: Insel and Young, 2001; Insel, 2010.

17. **Oxytocin and in-group/out-group dynamics**: De Dreu et al., 2011.

Chapter sources: This chapter draws strongly from the "Neuroscience and Psychology" section of the bibliography. Key work can be found in the extensive psychology studies of Daniel Batson (2010, 2014) and in the neuroscience studies of Tania Singer (Seymour et al., 2007; Singer, 2008; Tusche et al., 2016; Bosworth et al., 2016), as well as work by Bastiaansen et al.

(2009) and Marsh (2017). The chapter also built on the work by Howard Rachlin and others that humans have a social discounting function that is related to empathy (Jones and Rachlin, 2006; Rachlin and Jones, 2008a, 2008b; Locey et al., 2013; Marsh, 2017; Caldwell, 2017; Fernandes et al., 2020). See also the parochial altruism work (Wright, 1995, 1999; Turchin, 2003, 2006, 2016; Bregman, 2020; Hare and Woods, 2020; McCullough, 2020) but here note the importance of dehumanization therein (Bernard et al., 1965; Bandura et al., 1975; Gould, 1981; Grossman, 1995; Haidt, 2012; Hare and Woods, 2020; Wilkerson, 2020).

Note on emotion: Exactly what emotions are, how they are categorized, how similar they are across cultures, and how they vary across cultures is beyond the scope of this book. I refer readers to the several excellent popular neuroscience books on this question, such as Lisa Feldman Barrett's *How Emotions Are Made* (Barrett, 2017), Antonio Damasio's *Looking for Spinoza: Joy, Sorrow, and the Feeling Brain* (Damasio, 2003), Joe LeDoux's *The Emotional Brain* (LeDoux, 1996), and my own *The Mind within the Brain* (Redish, 2013b) for discussions on emotion, categorization, and the complexities therein. Jackson et al. (2019) is a fascinating study on how emotional categorization changes across cultures.

CHAPTER 11

1. **Fear-potentiated startle, rats:** Davis et al., 1993; Myers and Davis, 2002.

2. **PTSD symptoms:** Shiromani et al., 2009; Crocq and Crocq, 2000; Talarico and Rubin, 2003; Jones and Wessely, 2005.

3. **PTSD in ancient Assyria:** Abdul-Hamid and Hughes, 2014.

4. **PTSD and suicide:** Yehuda and Ledoux, 2007; Maguen et al., 2011; Bateson, 2015.

5. **Herodotus:** Herodotus, 430 BCE.

6. **Predictor of soldier's heart:** Pizarro et al., 2006.

7. **Hobos:** Crocq and Crocq, 2000; Jones and Wessely, 2005; Pizarro et al., 2006; *Frontline*, 2009.

8. **Regular economic crises:** Zarnowitz, 1992.

9. **Computational psychiatry:** Redish and Gordon, 2016; Anticevic and Murray, 2017; Series, 2020.

10. **Action-selection consequences of PTSD:** *Hyperalert:* Da Costa, 1884; Davis et al., 1993; Shay, 1994; Myers and Davis, 2002; Jones and Wessely, 2005. *Freeze/flee:* Roelofs, 2017; Hashemi et al., 2019. *Suicide:* Maguen et al., 2011; Bateson, 2015.

11. **Violent actions in some subjects with PTSD:** McFall et al., 1999; Glenn et al., 2002; Mac-Manus et al., 2015; Shay, 1994, 2002.

12. **Deliberative disruptions:** Shay, 2002; Bateson, 2015; Maguen et al., 2011.

13. **Kim Phuc:** N. Ut (1972), *The Napalm Girl*, photograph. See also the Kim Phuc interview on *Fresh Air with Terry Gross* (2001), NPR.

14. **PTSD sources:** Dunmore et al., 1999; Nutt, 2000; Heim et al., 2010.

15. **Civil War predictors of soldier's heart:** Pizarro et al., 2006.

16. **Bowlby and Harlow:** Bowlby, 1969; Harlow, 1986; see the history in Blum (2002).

17. **Chimpanzee grief:** Goodall, 1986, 1990; shown beautifully in the 2013 movie *Chimpanzee*.

18. **Neonatal Intensive Care Unit survivability:** Blum, 2002; Pados, 2019.

> **Chapter sources:** For clinical descriptions of PTSD: Kessler et al., 1995; Yehuda and Ledoux, 2007; Shiromani et al., 2009; Maguen et al., 2011; Ustinova and Cardeña, 2014. For historical descriptions of PTSD: Da Costa, 1884; Jones and Wessely, 2005; *Frontline*, 2009. Two of the best books on the difficulty of integrating back into society are from Jonathan Shay (1994, 2002). For clinical descriptions of resilience in humans: Bonanno et al., 2002; Hoge et al., 2007; Feder et al., 2010; Park, 2010; Horn and Feder, 2018. In nonhuman animals: Levine, 1957; Harlow, 1986; Parker and Maestripieri, 2011; Bergström et al., 2008. Keys to resilience: *Genetics and vulnerabilities:* Charney, 2004; Pitman et al., 2006; Gilbertson et al., 2002, 2010. *Learning from manageable stress:* Parker and Maestripieri, 2011; Russo et al., 2012; Horn and Feder, 2018; Kentner et al., 2019. *Social support:* Bowlby, 1969; Horn and Feder, 2018. *Meaning:* Feder et al., 2010; Park, 2010; Mahmoud and Rothenberger, 2019.

CHAPTER 12

1. **Moral rules beyond cooperation:** Simoons, 1994; Boyer, 2001; Atran, 2002; Tetlock, 2003; Haidt, 2012.

2. **Utilitarianism and harm avoidance:** Bentham, 1789/1907; Mill, 1861; Singer, 1975; Sinnott-Armstrong, 2009; Harris, 2010; Churchland, 2011, 2019; Greene, 2013.

3. **World is getting better:** Wright, 1999; Pinker, 2011; Rosling, 2018; McCullough, 2020.

4. **Evolution chasing a changing landscape:** Darwin, 1859; Gould, 1983; Wiener, 1995; Mayr, 2008.

5. **Parochial altruism:** Sober and Wilson, 1998; Wilson, 2010, 2015; Boehm, 2012; Haidt, 2012; Diamond, 2013; Fiske and Rai, 2015; Hare and Woods, 2020.

6. **Utilitarianism and neurophysiology:** Churchland, 2011, 2019; Greene, 2013.

7. **Haidt's moral matrix:** Haidt, 2012. There are a number of such attempts to describe how humans behave in the "moral realm." For example, Alan Fiske and Tage Rai have proposed a similar taxonomy based on four "relationship models": community sharing, authority ranking, equality matching, and market pricing (Rai and Fiske, 2011; Fiske and Rai, 2015). Community sharing is about doing things that encourage the formation of community, which can include doing horrible things to people outside of the community. Authority ranking is about creating, maintaining, and manipulating hierarchies. Equality matching is about ensuring balance between individuals through mechanisms like reciprocity and turn taking. And market pricing is about finding proportional valuations that can be exchanged.

8. **Haidt on Western bias:** Haidt, 2012, p. 120. As an alternate perspective, compare, for example, Isabel Wilkerson's (2020) discussion of the limitations and consequences of caste in India, the US, and Nazi Germany.

9. **Pinker on violence:** Pinker, 2011. Pinker also notes that there are dark sides that must be accommodated and addressed as well, and understanding the sources of violence are important steps to reduce them (see also Fiske and Rai, 2015). *Leviathan*, from Hobbes (1651). *Expanding Circle* and the *Escalator of Reason*, from Singer (1981).

10. **Empathy, moral structures, and altruism:** Allport, 1954; Batson, 2014; Nordell, 2017, 2021.

11. **Effects of walls:** Wapner, 2020.

12. **Doctors, ethics, and torture:** American Medical Association (2020), *AMA Principles of Medical Ethics*; American Psychological Association (2015), *APA's Council Bans Psychologist Participation in National Security Interrogations*, https://www.apa.org/news/press/releases/2015/08/psychologist-interrogations; see also *Economist* (July 28, 2015), How America's psychologists ended up endorsing torture, https://www.economist.com/democracy-in-america/2015/07/28/how-americas-psychologists-ended-up-endorsing-torture.

    Chapter sources: This chapter builds on recent work looking at the process of science (Laudan, 1978; Ben-Ari, 2011; Douglas, 2014; Firestein, 2012, 2015; Levenstein et al., 2020). Concepts of the social construction of our reality go back to Aristotle (350 BCE/2004) but draw more recently from Schelling (1978); Searle (1995); Enfield and Levinson (2006); and Young (2015), as well as from the group dynamics literature we've been building on (Wright, 1999; Hare and Woods, 2020; McCullough, 2020). The historical nature of descriptive and prescriptive issues of science go back to Hume (1777/1912); see also Gould (2000) and Dixon (2008). For completeness, one can look at historical views of the process of science from Popper (1959) and Kuhn (1962), but more modern views see the process of science differently.

CHAPTER 13

1. **American Civil War:** Foner, 2010. Of course, I do not want to suggest that the Union side in 1860 was providing equality. But it was providing a step in the right direction, particularly relative to what was being offered in the Confederacy. There are, of course, more steps to take to get to a true assurance game.

2. **Posture and submit:** Grossman, 1995.

3. **Evolutionary-built social interactions:** *In humans:* Tomasello, 2016; Marsh, 2017; Sapolsky, 2017. *In other primates:* Kummer, 1971; Goodall, 1986; de Waal, 1982, 1989; Strum, 1987; Cheney and Seyfarth, 1990; Sapolsky, 2001.

4. **Cap-and-trade:** Barreca et al., 2017; Schmalensee and Stavins, 2017.

5. **Tennessee house fire:** N. R. Nau (2013), Pay for spray fire protection policy: A case study of Obion County, Tennessee, *New Visions for Public Affairs* 5:41–54. See also C. Hendricks (2010), Fire chief responds to questions after home left to burn, *Heartland News*, October 6, 2010, https://www.kfvs12.com/story/13281481/fire-chief-responds-to-burning-questions-after-home-left-to-burn/.

6. **Americans have little savings:** US Federal Reserve (2019), *Report on the Economic Well-Being of US Households in 2018*.

7.   **EMTALA:** See the Emergency Medical Treatment and Active Labor Act (EMTALA). See also American College of Emergency Physicians, *EMTALA Fact Sheet*, www.acep.org.

8.   **Anatole France quote:** A. France (1894), *The Red Lily*.

9.   **Jefferson and Hemings:** Thomas Jefferson's relationship with Sally Hemings is fraught with complexity. She was his slave and therefore presumably unable to refuse his sexual advances. And yet her children and grandchildren described a more mutual relationship. Even his opponents described the relationship as "an affair." Hemings was the half-sister of his wife; Jefferson and Hemings began sleeping together after his wife died, likely while they were in Paris, where she was, ostensibly, legally a free woman although, presumably, without a lot of available resources. Yet she was sixteen when she arrived in Paris and nineteen when she left. He was forty-five and one of the most famous men in the world. And she became legally a slave again when they returned to Monticello, a journey she agreed to take, according to their children, only after a deal was made that the children would be free. He never did free Hemings herself, although it is unclear whether she would have been safer in that society if he had. He did free the children as per their agreement. Humans are fascinatingly complex in their relationships. See E. A. Foster et al. (1998), *Nature* 396:27–28 and A. Gordon-Reed (2008), *American Heritage* 58:5; as well as A. Gordon-Reed (1998), *Thomas Jefferson and Sally Hemings: An American Controversy*, University of Virginia Press; and www.monticello.org.

10.   **People expect some inequality:** Norton and Ariely, 2011.

11.   **Policing:** Horowitz, 2016; Bagwell et al., 1987–1990; Potter, 2013; Stoughton, 2016; Friedman, 2021; Delehanty et al., 2017; Mummolo, 2018; McElrath and Tuberville, 2020; Nordell, 2021; see US President's Task Force on 21st Century Policing (2015), *Final Report of the President's Task Force on 21st Century Policing*.

12.   **Modern views on addiction:** Leshner, 1997; Volkow and Li, 2004; Redish et al., 2008; Heyman, 2009; Hart, 2013.

13.   **Addiction and drug use are different:** Altman et al., 1996; Robbins and Clark, 2015; Schüll, 2012; Redish et al., 2008; Heyman, 2009; Hart, 2013.

14.   **Historically, addiction was seen as a moral failing:** Harding, 1986; Barry et al., 2014; Volkow, 2014. These views can be traced through literature: Robert Louis Stevenson's *The Strange Case of Dr. Jekyll and Mr. Hyde* (1886); the infamous movie *Reefer Madness* (1936); the movie *The Lost Weekend*, directed by Billy Wilder and written by Charles Brackett (1945); and Eugene O'Neill's play *Long Day's Journey into Night* (1956).

15.   **Opium wars:** Hanes and Sanello, 2002; Platt, 2018.

16.   **Addiction as disease/medical question:** Leshner, 1997; Meyer and Mirin, 1979; Volkow and Li, 2004; Koob and Le Moal, 2006.

17.   **Nixon and Reagan anti-drug-use speeches:** R. Nixon (1971), Drug abuse is "public enemy number one"; R. Reagan (1986), Remarks as president signs proclamation for "just say no" week.

18.   **Drug treatment better than jail:** Manski et al., 2001; Alexander, 2010; Leukefeld et al., 2011; DuVernay, 2016. See US Department of Justice (1992), *Drugs, Crime, and the Justice System*.

19. **Addiction as a brain disorder:** Meyer and Mirin, 1979; Leshner, 1997; Volkow and Li, 2004.

20. **Addiction as a disorder of decision-making systems:** Redish et al., 2008; Heyman, 2009; Redish, 2013b, 2020; Volkow and Li, 2004.

21. **Nicotine replacement therapy, methadone treatments:** Meyer and Mirin, 1979; Balfour and Fagerström, 1996; Benowitz, 1996; Hanson et al., 2003.

22. **Drug use shows elasticity:** Hursh, 1991; Grossman and Chaloupka, 1998; Bickel and Marsch, 2001; Liang and Chaloupka, 2002; Bruner and Johnson, 2013.

23. **Addiction as low elasticity:** Robinson, 2004; Rutledge, 2006; Schüll, 2012.

24. **Future is abstract and uncertain:** Ainslie, 1992; Trope and Liberman, 2003; Kurth-Nelson et al., 2012. *Making the future concrete reduces discounting:* Peters and Büchel, 2010, 2011. *Possible to train people to see future as concrete:* Radu et al., 2011; Stein et al., 2016; Snider et al., 2016.

25. **Contingency management:** Petry, 2012; Higgins et al., 1991; Lussier et al., 2006.

26. **Contingency management works better than it should:** Lussier et al., 2006; Regier and Redish, 2015; Hart, 2013.

27. **Contingency management as shifting decision systems:** Regier and Redish, 2015.

28. **Default options:** *Retirement plans:* Thaler and Benartzi, 2007. *Organ donation:* Davidai et al., 2012.

29. **Hollywood and the tobacco industry:** Lum et al., 2008; Mekemson and Glantz, 2002.

30. **Possession increases value:** Bray et al., 2008; Wang, 2009; Thaler and Sunstein, 2008; Thaler, 2015; Akerlof and Shiller, 2015; Kahneman, 2011.

31. **Anti-government, anti-tax groups:** Rothbard, 2002; Fontinelle, 2020.

32. **Perspective taking:** Pinker, 2011; Nordell, 2017, 2021. See also chapter 10.

Chapter sources: The ideas in this chapter draw from work by Schelling (1978), Searle (1995, 2006), and Thaler and Sunstein (2008; see also Thaler, 2015). The discussion of how rules can be used to empower one group over another drew ideas experimentally from Axelrod (1986; Axelrod and Dion, 1988) and sociologically from Alexander (2010); Coates (2015); and Wilkerson (2020). The discussion of addiction draws heavily on my own work (Redish et al., 2008; Redish, 2013b, 2020), as well as work by Hart (2013, 2017); Heyman (2009); Goldstein (2000); and others as noted above. The discussion of self-regulation and self-control draw from arguments and controversies between Baumeister et al. (1994, 2007); Rachlin (2000); and others (Ainslie, 1992, 2001; Crockett et al., 2013; Bechara, 2005). See also my lab's work on issues of self-control (Kurth-Nelson and Redish, 2010, 2012; Redish, 2013b).

CHAPTER 14

1. **Felons and reputation markers:** Alexander, 2010.

2. **Pavlovian desire for retributive justice:** Hoffman et al., 1994.

3. **Catholic Church and the pedophilia cover-up:** Spotlight Team (2002), *Boston Globe*, www3 .bostonglobe.com/arts/movies/spotlight-movie.

4. **Extinguishing responses:** Pavlov, 1927; Bouton, 2007; Redish et al., 2007.

5. **1994 crime bill:** J. Gramlich (October 17, 2019), Five facts about crime in the US, Pew Research Center, https://www.pewresearch.org/fact-tank/2019/10/17/facts-about-crime-in -the-u-s/; J. Pfaff (April 12, 2016), Bill Clinton is wrong about his crime bill. So are the protesters he lectured, *New York Times*, https://www.nytimes.com/2016/04/12/magazine /bill-clinton-is-wrong-about-his-crime-bill-so-are-the-protesters-he-lectured.html.

6. **Private prisons:** Bauer, 2018; DuVernay, 2016.

7. **Prisons and recidivism:** US Department of Justice (1992), *Drugs, Crime, and the Justice System*. See also Bregman, 2020.

8. **Punishments start small:** Acheson, 1988; Ellickson, 1994; Ostrom et al., 1994; Boehm, 2012; Wilson, 2010, 2015.

9. **Poor estimates of wrongs done:** Ellickson, 1994; Hoffman et al., 1994; Haidt, 2012; Anderson et al., 2020; Crockett, 2017.

10. **Restorative justice:** Braithwaite, 2004; Gavrielides, 2005; Boehm, 2012; Diamond, 2013.

11. **Trustee game:** King-Casas et al., 2008.

12. **Discussions of reparations in the US:** Alexander, 2010; Coates, 2014; Wilkerson, 2020.

13. **Complexities in reparations:** *Planet Money* (2020), Reparations in New Zealand, NPR; National Library of Israel (2020), The Reparations Agreement of 1952 and the response in Israel, https://web.nli.org.il/sites/NLI/English/collections/personalsites/Israel-Germany /Division-of-Germany/Pages/Reparations-Agreement.aspx.

14. **German reparations for the Holocaust:** National Library of Israel (2020), The Reparations Agreement of 1952 and the response in Israel; *Times of Israel* (2012), Germany increases reparations for Holocaust survivors; *Haaretz* (2018), Germany increases funding for Holocaust survivors by $88M. In 2019 alone, Germany paid over $500 million to the Claims Conference (the negotiating team that meets annually).

15. **Punishment in tribal groups:** Boehm, 2012. See also Diamond (2013) for similar descriptions from other hunting/gathering/foraging populations.

16. **Emotional guilt is larger if you got away with it:** Katchadourian, 2011.

17. **John Lewis:** Lewis, 1998, 2012.

18. **Graded punishments:** *California cattle ranchers:* Ellickson, 1994. *Fisherman and farmers:* Acheson, 1988; Ostrom, 1990; Burger et al., 2001. *Religious communities:* Wilson, 2010.

19. *Erewhon:* Butler's 1872 novel was written as Victorian satire, treating crime as a disease and disease as a crime. However, there are discussions today about whether justice should be medicalized rather than delivered as punishment (Greene and Cohen, 2004; Jones et al., 2014).

20. **Neurolaw concerned about blame given physical change**: Greene and Cohen, 2004; Jones et al., 2014; Slobogin, 2017; Pernu and Elzein, 2020.

21. **Case study of brain tumor and pedophilia**: Burns and Swerdlow, 2003; Batts, 2009; Jones et al., 2014.

    Chapter sources: This chapter draws from Hoffman et al. (1994), Sandel (2009), and the new work on neurolaw by Jones et al. (2014). Boehm (2012) and Diamond (2013) have good discussions on justice questions in the context of smaller communities. I started working out some of the relationships between justice and decision systems in an earlier paper (Redish, 2013a). Katchadourian (2011) has a fascinating study on the concept of "guilt." The chapter also draws from the structures of community and the first-, second-, and third-party social control that we explored in the first part of the book (chapters 4–6) and from ideas on how that expands into larger communities (Ostrom, 1990; Ellickson, 1994; Wright, 1999; Wilson, 2010, 2015; McCullough, 2020).

CHAPTER 15

1. **Atheists are moral**: Anderson and Mellor, 2009; Grossman and Parrett, 2011. See also the discussion of aspirational goals from nonreligious sources, such as the US Declaration of Independence (1776), the French *Déclaration des droit de l'homme et du citoyen* [Declaration of the Rights of Man and of the Citizen] (1789), or the United Nation's Universal Declaration of Human Rights (1948).

2. **Initiations drive loyalty**: Aronson and Mills, 1959; Sosis and Bressler, 2003; Gerard and Mathewson, 1966; Inzlicht et al., 2018.

3. *Tikkun olam*: Talmud, Pirkei Avot 2:16.

4. **Immutable absolutes**: Tetlock et al., 2000; Tetlock, 2003; Pinker, 2011; Haidt, 2012.

5. **Willpower takes effort**: Baumeister et al., 1994; Rachlin, 2000.

6. **Purity**: Haidt, 2012; see also Atran, 2002.

7. **Pavlovian learning**: Pavlov, 1927; Domjan, 1998; Bouton, 2007; Redish, 2013b.

8. **Impure often seen as inhuman**: Turchin, 2003; Hare and Woods, 2020.

9. **Food taboos**: Simoons, 1994; Boyer, 2001; Atran, 2002.

10. **Moral judgments**: Tetlock et al., 2000; Tetlock, 2003; Berman and Small, 2018; Anderson et al., 2020.

11. **Arbitrary rules as community**: Iannaccone, 1992; Boyer, 2001; Wilson, 2010, but see Atran (2002) for an alternate view.

12. **Islamic finance**: H. Osborne (October 29, 2013), Islamic finance, *Guardian*, https://www .theguardian.com/money/2013/oct/29/islamic-finance-sharia-compliant-money-interest.

13. **Jubilee years**: Graeber, 2014.

14. **Catholic Church and the pedophilia scandal:** Spotlight Team (2002), *Boston Globe*, www3 .bostonglobe.com/arts/movies/spotlight-movie.

15. **Ninth commandment is about gossip:** Wilson, 2010; Boehm, 2012.

Chapter sources: This chapter builds on analyses of religion from the perspective of psychology, sociology, and mesoeconomics (Atran, 2002; Boyer, 2001; Iannaccone, 1992; Tetlock et al., 2000; Tetlock, 2003; Wilson, 2010; Haidt, 2012; Anderson et al., 2020; Hare and Woods, 2020), as well as older work on universalities of religion from a humanist perspective (Campbell, 1959–1968; Levi-Strauss, 1970; Vico, 1725). Sinnott-Armstrong (2009) has an interesting deconstructionist view on the relationship between religion and morality.

CHAPTER 16

1. **Social norms are often implicit:** Levi-Strauss, 1970; Boehm, 2012; Diamond, 2013; Henrich, 2016.

2. **"Governments are consensus fictions pretended into place"** is a quote from Lois McMaster Bujold's novel *Barrayar* (1991), Baen Books, p. 48.

3. **Athenian citizen responsibilities:** Thucydides, 403 BCE; Herodotus, 430 BCE; Stone, 1989.

4. **Paul in jail:** Acts 22.

5. **Purity and power:** Wilkerson, 2020; Haidt, 2012; Harari, 2015; McGhee, 2021.

6. **"A republic, if you can keep it"**: see Papers of Dr. James McHenry on the Federal Convention of 1787, *American Historical Review* 11(1906):595–624.

7. **Journeys driven by appeal to aspirational goals:** *Thirteenth Amendment:* Foner, 2010; *Civil Rights and Voting Rights Acts:* Caro, 2013; Lewis, 2012; *Gender equality:* B. West and J. Cohen (2018), *R. B. G.*, Magnolia Pictures.

8. **India trading companies:** Durant, 1930; Hanes and Sanello, 2002; Platt, 2018; see, for example, *Charter of the Dutch West India Company* (1629), Avalon Project, Yale Law School, avalon.law.yale.edu/17th_century/westind.asp.

9. **Greek debt:** Graeber, 2014. See also Greece's debt, Council on Foreign Relations, www.cfr .org/timeline/greeces-debt-crisis-timeline.

10. **Profit as corporate aspirational goals:** Friedman, 1970; Fox, 2009; Stout, 2012. Lynn Stout points out in her 2012 book *The Shareholder Value Myth* that the short-term profit motive is neither a legal rule nor necessarily the best motive available to the corporation.

11. **US Postal Service:** Gallagher, 2017.

Chapter sources: This chapter builds on extensive work in the anthropology, sociology, and ethology literatures (Turchin, 2003, 2006, 2016; Levi-Strauss, 1970; Wright, 1999; Boehm, 2012; Harari, 2015; McCullough, 2020), as well as some of the experimental mesoeconomics literature (Runge, 1981; Ostrom et al., 1994; Wilson, 2015). The moral structural changes in the US drew on historical documents (Foner, 2010; Caro, 2013), as well as more recent sociological perspectives on race and history (Lewis, 2012; Wilkerson, 2020; Coates, 2015; DuVernay, 2016; Alexander, 2010; McGhee, 2021).

CHAPTER 17

1. **Oxytocin:** Insel, 2000; Insel and Young, 2001; MacDonald and MacDonald, 2010; Neto et al., 2020. Also see chapter 10.

2. **Integration increases tolerance:** Allport, 1954; Nordell, 2017, 2021.

3. **Celts, Romans, and Hannibal:** O'Connell, 2010; Turchin, 2003.

4. **"We need an alien invasion":** Paul Krugman (2011), *Fareed Zakaria GPS*, August 14; see also the *Watchmen* 1987 graphic novel and 2009 movie.

5. **Globalization has increased nonzero productivity and reduced violence:** Wright, 1999; Pinker, 2011; McCullough, 2020.

6. **Animal welfare history:** Singer, 1975; Blum, 1995; see the UK Cruelty to Animals Act (1835).

7. **Animal cognition:** *Primate politics:* Goodall, 1986; de Waal, 1982, 1989; Strum, 1987; Cheney and Seyfarth, 1990. *Dogs and cats:* Hare and Woods, 2020. *Rats can plan:* Johnson and Redish, 2007; Redish, 2016.

8. **Octopus intelligence:** Godfrey-Smith, 2016.

9. **Pet statistics:** American Veterinary Medical Association (2018), *2017–2018 US Pet Ownership and Demographics Sourcebook.*

10. **Oxytocin and pets:** Feldman, 2012; Rilling, 2013.

11. **Service animals:** Americans with Disabilities Act, https://www.ada.gov/service_animals_2010.

12. **Emotional-support animals:** M. Pitofsky (October 10, 2018), Flight delayed after woman brings "emotional support squirrel" on plane, *USA Today*; D. Silva (January 30, 2018), Emotional support peacock denied flight by United Airlines, NBC News, https://www.nbcnews.com/storyline/airplane-mode/emotional-support-peacock-denied-flight-united-airlines-n842971; F. Street (October 23, 2019), Should emotional-support animals be allowed on planes?, CNN, https://www.cnn.com/travel/article/emotional-support-animals-airplanes/index.html; see US Department of Transportation, www.transportation.gov/individuals/aviation-consumer-protection/service-animals-including-emotional-support-animals.

13. **Biodiversity increases ecosystem productivity:** Loreau et al., 2001; Tilman et al., 2014.

14. **Destabilizing civilizations:** Turco et al., 1983; see also Diamond, 2005.

15. **Earthrise:** National Aeronautics and Space Administration, NASA image AS08-14-2383, www.nasa.gov/multimedia/imagegallery/image_feature_1249.html.

16. **Ridley Scott on *Blade Runner*:** Ridley Scott interview on *Empire Film Podcast*, episode 283.

Chapter sources: This chapter expands on the questions of parochial altruism that we have been discussing throughout the book, including the idea that a key question is who to include in the assurance game (Turchin, 2003; Sinnott-Armstrong, 2009; Diamond, 2013; Hare and Woods, 2020; Wilkerson, 2020) and the idea that the story of humanity is a continuous shift to larger and larger communities (Wright, 1999; Turchin, 2003, 2016; Pinker, 2011; Diamond, 2013; Harari, 2015; McCullough, 2020).

Note on communities: There are excellent nonfiction books making the case that we are developing larger and larger communities based on expanding scales (Singer, 1981; Wright, 1999; Rachlin and Jones, 2008a; Pinker, 2011; Wilson, 2015; Turchin, 2016; Marsh, 2017; McCullough, 2020); however, I have to admit that the scaling sequence described here derives from Orson Scott Card's fictional humanist philosopher Valentine Wiggin from his novel *Speaker for the Dead*. Valentine (who uses the internet handle Demosthenes) identifies five terms of expanding otherness: *utlänning* (otherlander), *främling* (stranger), *raman* (sentient alien), *varelse* (true alien), and *djur* (beast). Someone who is *utlänning* is from our group but from another village, so while we share ethnicity and culture, we may well still be in conflict with them. Someone who is *främling* is from another world, so we recognize them as human but find their ethnicity and culture different and off-putting. Someone who is *raman* is alien to us. We may recognize them as sentient and intelligent but do not understand them or their motivations. Someone who is *varelse* is too alien to even recognize as sentient. And finally, *djur* is the beast, the evil that must be destroyed. I have added *kith and kin* as a sixth (first) level before *utlänning*.

CHAPTER 18

1. **Vaccines do not cause autism:** B. Taylor et al. (1999), *Lancet* 353:2026–2029; Murch et al. (2004), *Lancet* 363:750; R. Horton (2004), *Lancet* 363:820–821; Godlee et al. (2011), *British Medical Journal* 342:c7452; B. Deer (2011), *British Medical Journal* 342:c5347.

2. **Better for the individual to be vaccinated:** Some individuals with specific allergies and other conditions cannot receive vaccinations, but these individuals have well-identified health issues. For people without these conditions, vaccination is not only generally safe but protective and much safer than getting the disease. Because some people cannot get vaccinated, it is important for others to do so to create herd immunity and prevent diseases from taking hold in the population. See Centers for Disease Control and Prevention, www.cdc.gov/vaccines/vpd/should-not-vacc.html.

3. **Pecuniary rewards can backfire:** Gneezy and Rustichini, 2000; Bosworth et al., 2016; Bowles, 2016; see chapter 6.

4. **Two stable conditions:** Marwell and Ames, 1981; Runge, 1981, 1984b; Fehr and Fischbacher, 2004; Greene, 2013; Henrich, 2016.

5. **Games bifurcate:** Taylor, 1976; Marwell and Ames, 1981; Berg et al., 1995; Fehr and Fischbacher, 2004; King-Casas et al., 2005; Bowles and Gintis, 2011; Bowles, 2016; Alos-Ferrer and Farolfi, 2019; Henrich and Muthukrishna, 2021; van Dijk and De Dreu, 2021.

6. **Fischbacher, Gächter, and Fehr study:** Fischbacher et al., 2001; Fehr and Fischbacher, 2004.

7. **Cultural differences in public goods games:** Greene, 2013.

8. **Evolutionary niches depend on others:** Maynard-Smith, 1982; Mayr, 2008.

9. **Laffer curve:** Wanniski, 1978; Shiller, 2017; Trabandt and Uhlig, 2011. These analyses only apply to governments that do not issue their own currency. Taxes play a different role in governments that issue their own currency. These macroeconomic mathematics are beyond the scope of this book, but I refer the reader to Kelton (2020) for a thorough discussion of them.

10. **Social capital:** Putnam, 2000; Atran, 2002; Diamond, 2005; Wapner, 2020, Putnam and Garrett, 2020.

11. **Thatcher quote:** M. Thatcher (1987), Margaret Thatcher Foundation, www.margaretthatcher.org/document/106689.

12. **Craig Nelson quote:** Craig T. Nelson, comments made to *Glenn Beck*. See *The Daily Show with Jon Stewart*'s A moment of Zen, www.cc.com/video-clips/thzt6m.

CHAPTER 19

1. **Interrelated networks reduce conflict:** Pinker, 2011.

2. **People's valuation of membership depends on cost:** Aronson and Mills, 1959; Gerard and Mathewson, 1966; Sosis and Bressler, 2003; Inzlicht et al., 2018.

3. **Rituals:** McNeill, 1995; Wright, 1999; Atran, 2002.

4. **Cold War defections:** Carruthers, 2005.

5. **Parochial altruism:** Turchin, 2003, 2016; Wilson, 2010, 2015; Greene, 2013; Henrich, 2016; Tomasello, 2016; Diamond, 1997, 2005; Boehm, 2012; Hare and Woods, 2020.

6. **Nested groups in anthropological studies:** Levi-Strauss, 1970; Diamond, 1997; Wright, 1999; Turchin, 2003; Boehm, 2012. *Nested groups in modern societies:* Wright, 1999; Jones and Rachlin, 2006; Pinker, 2011; Turchin, 2016; McCullough, 2020.

7. **Bobsledding:** N. Atkin (February 5, 2014), The real "Cool Runnings," *ESPN*, http://en.espn.co.uk/olympic-sports/sport/story/280229.html; S. Bono (February 15, 2018), The real "Cool Runnings": 1988 Jamaican bobsledder says true story was "Even more remarkable," *Inside Edition: Sports*, https://www.insideedition.com/real-cool-runnings-1988-jamaican-bobsledder-says-true-story-was-even-more-remarkable-40754.

8. **Tangled webs reduce conflict:** Pinker, 2011; see also Ostrom et al., 1994; Wilson, 2015; Fiske and Rai, 2015; Putnam and Garrett, 2020.

9. **Coordination has increased through changes in physical technologies:** Wright, 1999; Pinker, 2011, Henrich, 2016; McCullough, 2020.

10. **Small correlations swamped by diversity and other factors:** Gould, 1981; Fine, 2010.

11. **Preferred levels of inequality:** Norton and Ariely, 2011.

12. **Children and fairness:** Tomasello, 2016.

13. **Salaries shift as women enter jobs:** Levanon et al., 2009.

14. **Safety nets increase fairness:** Stuckler and Basu, 2013.

15. **Humans evolved to understand complex others as agents and applied it to themselves:** Byrne and Whiten, 1989; Kurzban, 2010; Redish 2013b; Graziano, 2015; Gazzaniga, 2019.

16. **Peaceful protests work:** Chenowith and Stephan, 2011.

17. **Reparations:** T. Coates (June 2014), The case for reparations, *Atlantic*, https://www .theatlantic.com/magazine/archive/2014/06/the-case-for-reparations/361631/; *Planet Money* (2020), Reparations in New Zealand, NPR; R. Triesman (March 23, 2021), In likely first, Chicago suburb of Evanston approves reparations, NPR, https://www.npr.org/2021/03 /23/980277688.

CHAPTER 20

1. **Masks work to prevent COVID-19 transmission:** N. Leung et al. (2020), *Nature Medicine* 26:676–680; C. T. Leffler et al. (2020), *American Journal of Tropical Medicine and hygiene* 103:2400–2411; J. Abaluck et al. (2021), *Science* 375:eabi9069.

2. **American COVID-19 death toll:** Centers for Disease Control and Prevention, covid.cdc .gov/covid-data-tracker; Johns Hopkins University of Medicine, coronavirus.jhu.edu/data /mortality.

3. **Climate change is intergenerational:** Lohse and Waichman, 2020. See also the work on common pool resources (Runge, 1981, 1984a; Ostrom, 1990; Ostrom et al., 1994; Wilson, 2015) and the difficulty in playing a multigenerational game (Binmore, 2005).

4. **Moral arc bends toward justice:** Martin Luther King Jr. (1956), "Statement on ending the bus boycott"; Martin Luther King Jr. (1968), speech delivered at the National Cathedral, Washington, DC.

5. **Bistability:** Dawes, 1980; Marwell and Ames, 1981; Roth and Schoumaker, 1983; Ostrom et al., 1992; Fehr and Fischbacher, 2004; Bowles and Gintis, 2011.

6. **Subgroups of cooperators:** Turchin, 2003; Diamond, 2005.

Bibliography

SELECTED WORKS ON THE ECONOMICS
OF COMMUNITIES (MESOECONOMICS)

Alos-Ferrer, C., and Farolfi, F. (2019). Trust games and beyond. *Frontiers in Neuroscience* 13:887.

Axelrod, R. (1984). *The Evolution of Cooperation*. Basic Books.

Axelrod, R. (1986). An evolutionary approach to norms. *American Political Science Review* 80(4):1095–1111.

Axelrod, R., and Dion, D. (1988). The further evolution of cooperation. *Science* 242(4884):1385–1390.

Binmore, K. (2005). *Natural Justice*. Oxford University Press.

Bowles, S. (2016). *The Moral Economy*. Yale University Press.

Bowles, S., and Gintis, H. (2011). *A Cooperative Species: Human Reciprocity and Its Evolution*. Princeton University Press.

Camerer, C. F. (2003). *Behavioral Game Theory: Experiments in Strategic Interaction*. Princeton University Press.

Dawes, R. M. (1980). Social dilemmas. *Annual Review of Psychology* 31:169–193.

Fehr, E., and Fischbacher, U. (2004). Social norms and human cooperation. *Trends in Cognitive Sciences* 8(4):185–190.

Joyce, R. (2007). *The Evolution of Morality*. MIT Press.

Kollock, P. (1998). Social dilemmas: The anatomy of cooperation. *Annual Review of Sociology* 24:183–214.

Maynard-Smith, J. (1982). *Evolution and the Theory of Games*. Cambridge University Press.

McElreath, R., and Boyd, R. (2007). *Mathematical Models of Social Evolution: A Guide for the Perplexed*. University of Chicago Press.

Ostrom, E. (1990). *Governing the Commons: The Evolution of Institutions for Collective Action*. Cambridge University Press.

Poundstone, W. (1992). *Prisoner's Dilemma*. Anchor.

Runge, C. F. (1984a). Institutions and the free rider: The assurance problem in collective action. *Journal of Politics* 46(1):154–181.

Schelling, T. (1960). *The Strategy of Conflict*. Harvard University Press.

Schelling, T. (1978). *Micromotives and Macrobehavior*. W. W. Norton.

Skyrms, B. (2004). *The Stag Hunt and the Evolution of Social Structure*. Cambridge University Press.

Sober, E., and Wilson, D. S. (1998). *Unto Others: The Evolution and Psychology of Unselfish Behavior*. Harvard University Press.

Sugden, R. (1986). *The Economics of Rights, Co-operation and Welfare*. Springer.

Taylor, M. (1976). *Anarchy and Cooperation*. Wiley.

Wilson, D. S. (2015). *Does Altruism Exist?* Yale University Press.

Wright, R. (1999). *Nonzero: The Logic of Human Destiny*. Pantheon Books.

SELECTED WORKS ON DECISION-MAKING (NEUROSCIENCE AND PSYCHOLOGY)

Ainslie, G. (1992). *Picoeconomics*. Cambridge University Press.

Ainslie, G. (2001). *Breakdown of Will*. Cambridge University Press.

Barrett, L. F. (2017). *How Emotions Are Made: The Secret Life of the Brain*. Macmillan.

Batson, D. (2014). *The Altruism Question—toward a Social Psychological Answer*. Psychology Press.

Bowlby, J. (1969). *Attachment: Attachment and Loss*. Vol. 1. Basic Books.

Crockett, M. J. (2017). Models of morality. *Trends in Cognitive Sciences* 17(8):363–366.

Damasio, A. (1994). *Descartes' Error: Emotion, Reason, and the Human Brain*. Quill Press.

Damasio, A. (2003). *Looking for Spinoza: Joy, Sorrow, and the Feeling Brain*. Mariner.

Eagleman, D. (2011). *Incognito: The Secret Lives of the Brain*. Vintage Books.

Gazzaniga, M. S. (2012). *Who's in Charge? Free Will and the Science of the Brain*. Little, Brown.

Gazzaniga, M. S. (2019). *The Consciousness Instinct: Unraveling the Mystery of How the Brain Makes the Mind*. Farrar, Straus and Giroux.

Graziano, M. (2015). *Consciousness and the Social Brain*. Oxford University Press.

Greene, J. (2013). *Moral Tribes: Emotion, Reason, and the Gap between Us and Them*. Penguin.

Haidt, J. (2006). *The Happiness Hypothesis*. Basic Books.

Kahneman, D. (2011). *Thinking, Fast, and Slow*. Farrar, Straus and Giroux.

Katchadourian, H. (2011). *Guilt: The Bite of Conscience*. Stanford University Press.

Klein, G. (1999). *Sources of Power: How People Make Decisions*. MIT Press.

Kosslyn, S. M. (1994). *Image and Brain*. MIT Press.

Kurzban, R. (2010). *Why Everyone (Else) Is a Hypocrite*. Princeton University Press.

LeDoux, J. (1996). *The Emotional Brain: The Mysterious Underpinnings of Emotional Life*. Simon and Schuster.

Marsh, A. (2017). *The Fear Factor*. Basic Books.

Milgram, S. (1974). *Obedience to Authority*. Harper Perennial.

Rachlin, H. (2000). *The Science of Self-Control*. Harvard University Press.

Rai, T. S., and Fiske, A. P. (2011). Moral psychology is relationship regulation: Moral motives for unity, hierarchy, equality, and proportionality. *Psychological Review* 118(1):57.

Redish, A. D. (2013b). *The Mind within the Brain: How We Make Decisions and How Those Decisions Go Wrong*. Oxford University Press.

Thaler, R. H. (2015). *Misbehaving*. W. W. Norton.

Thaler, R. H., and Sunstein, C. (2008). *Nudge*. Yale University Press.

SELECTED WORKS ON ETHOLOGY, ANTHROPOLOGY, AND THE EVOLUTION OF BEHAVIOR

Alexander, R. D. (1974). The evolution of social behavior. *Annual Reviews Ecological Systems* 5:325–383.

Boehm, C. (2012). *Moral Origins: The Evolution of Virtue, Altruism, and Shame*. Basic Books.

Boesch, C., and Boesch-Achermann, H. (2012). *The Chimpanzees of the Taï Forest*. Oxford University Press.

Byrne, R. W., and Whiten, A., eds. (1989). *Machiavellian Intelligence: Social Expertise and the Evolution of Intellect in Monkeys, Apes, and Humans*. Oxford University Press.

Cheney, D. L., and Seyfarth, R. M. (1990). *How Monkeys See the World*. University of Chicago Press.

Darwin, C. (1859). *On the Origin of Species*. Broadview Press.

Dawkins, R. (1976). *The Selfish Gene*. Oxford University Press.

de Waal, F. (1982). *Chimpanzee Politics: Power and Sex among Apes*. Johns Hopkins University Press.

de Waal, F. (1989). *Peacemaking among Primates*. Harvard University Press.

de Waal, F. (2001). *The Ape and the Sushi Master: Cultural Reflections of a Primatologist*. Basic Books.

de Waal, F. (2009). *The Age of Empathy*. Random House.

de Waal, F., and Lanting, F. (1997). *Bonobo: The Forgotten Ape*. University of California Press.

Diamond, J. (2006). *The Third Chimpanzee*. HarperCollins.

Diamond, J. (2013). *The World until Yesterday*. Penguin.

Fossey, D. (1983). *Gorillas in the Mist*. Houghton Mifflin.

Goodall, J. (1990). *Through a Window*. Houghton Mifflin.

Goodall, J., and van Lawick, H. (1971). *In the Shadow of Man*. Houghton Mifflin.

Hare, B., and Woods, V. (2020). *Survival of the Friendliest*. Random House.

Henrich, J. (2016). *The Secret of Our Success*. Princeton University Press.

Hrdy, S. B. (1999). *Mother Nature: Maternal Instincts and How They Shape the Human Species*. Ballantine Books.

Kummer, H. (1971). *Primate Societies*. Aldine Atherton.

Sapolsky, R. M. (2001). *A Primate's Memoir*. Touchstone.

Sapolsky, R. M. (2017). *Behave*. Penguin.

Strum, S. (1987). *Almost Human*. Random House.

Tomasello, M. (2016). *A Natural History of Morality*. Harvard University Press.

Wright, R. (1995). *The Moral Animal: Why We Are the Way We Are: The New Science of Evolutionary Psychology*. Vintage Books.

SELECTED WORKS ON SOCIOLOGY AND THE CONSTRUCTION OF SOCIETIES

Acheson, J. (1988). *The Lobster Gangs of Maine*. University Press of New England.

Alexander, M. (2010). *The New Jim Crow: Mass Incarceration in the Age of Colorblindness*. New Press.

Allport, G. (1954). *The Nature of Prejudice*. Perseus Books.

Atran, S. (2002). *In Gods We Trust*. Oxford University Press.

Bagwell, O., et al., dirs. (1987–1990). *Eyes on the Prize: America's Civil Rights Movement*. 14 episodes. PBS.

Boyer, P. (2001). *Religion Explained*. Basic Books.

Bregman, R. (2017). *Utopia for Realists*. Little, Brown.

Bregman, R. (2020). *Humankind: A Hopeful History*. Little, Brown.

Chenowith, E., and Stephan, M. (2011). *Why Civil Resistance Works: The Strategic Logic of Nonviolent Conflict*. Columbia University Press.

Cronk, L., and Leech, B. L. (2013). *Meeting at Grand Central*. Princeton University Press.

Diamond, J. (1997). *Guns, Germs, and Steel*. W. W. Norton.

Diamond, J. (2005). *Collapse*. Viking Press.

DuVernay, A., dir. (2016). *13th*. Netflix.

Ellickson, R. C. (1994). *Order without Law*. Harvard University Press.

Enfield, N. J., and Levinson, S. C., eds. (2006). *Roots of Human Sociality: Culture, Cognition and Interaction*. Berg.

Haidt, J. (2012). *The Righteous Mind: Why Good People Are Divided by Politics and Religion*. Vintage Books.

McCullough, M. E. (2020). *The Kindness of Strangers: How a Selfish Ape Invented a New Moral Code*. Basic Books.

McGhee, H. (2021). *The Sum of Us: What Racism Costs Everyone and How We Can Prosper Together*. One World Press.

Nordell, J. (2021). *The End of Bias: A Beginning*. Metropolitan Books/Macmillan.

Pinker, S. (2011). *The Better Angels of Our Nature*. Penguin.

Putnam, R. D. (2000). *Bowling Alone: The Collapse and Revival of American Community*. Simon and Schuster.

Putnam, R. D., and Garrett, S. R. (2020). *The Upswing: How Americans Came Together a Century Ago and How We Can Do It Again*. Simon and Schuster.

Rosling, H. (2018). *Factfulness: Ten Reasons We're Wrong about the World—and Why Things Are Better Than You Think*. Flatiron Books.

Searle, J. R. (1995). *The Construction of Social Reality*. Free Press.

Stout, L. (2011). *Cultivating Conscience*. Princeton University Press.

Stout, L. (2012). *The Shareholder Value Myth*. Berrett-Koehler.

Stuckler, D., and Basu, S. (2013). *The Body Economic: Why Austerity Kills*. Basic Books.

Turchin, P. (2003). *Historical Dynamics*. Princeton University Press.

Turchin, P. (2006). *War and Peace and War*. Penguin.

Turchin, P. (2016). *Ultrasociety*. Beresta Books.

Wilkerson, I. (2020). *Caste: The Origins of Our Discontents*. Random House.

Wilson, D. S. (2010). *Darwin's Cathedral: Evolution, Religion, and the Nature of Society*. University of Chicago Press.

SELECTED WORKS ON MORAL PHILOSOPHY AND LEGAL STUDIES

Anderson, R. A., Crockett, M. J., and Pizarro, D. A. (2020). A theory of moral praise. *Trends in Cognitive Sciences* 24(9):694–703.

Aristotle. (350 BCE/1925). *Nichomachean Ethics*. Translated by D. Ross. Oxford University Press.

Aristotle. (350 BCE/2004). *Politics: A Treatise on Government*. Translated by W. Ellis, 1912. Project Gutenberg.

Churchland, P. S. (2011). *Brain Trust*. Princeton University Press.

Churchland, P. S. (2019). *Conscience*. W. W. Norton.

de Waal, F. B. M. (2014). Natural normativity: The "is" and "ought" of animal behavior. *Behaviour* 151:185–204.

Edmonds, D. (2015). *Would You Kill the Fat Man?* Princeton University Press.

Greene, J., and Cohen, J. (2004). For the law, neuroscience changes nothing and everything. *Philosophical Transactions of the Royal Society B* 359:1775–1785.

Harris, S. (2010). *The Moral Landscape*. Free Press.

Hobbes, T. (1651). *Leviathan: Or the Matter, Forme and Power of a Commonwealth, Ecclesiasticall and Civil*. Printed by the author.

Hoffman, M. B. (2014). *The Punisher's Brain*. Cambridge University Press.

Hume, D. (1777/1912). *An Enquiry Concerning the Principles of Morals*. Project Gutenberg.

Jones, O., Shen, F. X., and Schall, J. D. (2014). *Law and Neuroscience*. Wolters Kluwer.

Kant, I. (1781/1787). *Critique of Pure Reason*. Printed by the author.

Locke, J. (1690). *Two Treatises on Government*. Awnsham Churchill.

Mill, J. S. (1861). *Utilitarianism*. Longmans, Green.

Rawls, J. (1971). *A Theory of Justice*. Harvard University Press.

Rawls, J. (1985). Justice as fairness: Political not metaphysical. *Philosophy and Public Affairs* 14(3):223–251.

Rousseau, J. J. (1762). *The Social Contract*. Marc Michel Rey.

Sandel, M. (2009). *Justice: What's the Right Thing to Do?* Farrar, Straus and Giroux.

Scanlon, T. (1998). *What We Owe to Each Other*. Harvard University Press.

Shermer, M. (2004). *The Science of Good and Evil*. Henry Holt.

Shermer, M. (2015). *The Moral Arc: How Science and Reason Lead Humanity toward Truth, Justice, and Freedom*. Macmillan.

Sinnott-Armstrong, W. (2009). *Morality Without God?* Oxford University Press.

SELECTED WORKS ON THE PHILOSOPHY OF SCIENCE

Ben-Ari, M. (2011). *Just A Theory: Exploring the Nature of Science*. Prometheus Books.

Douglas, H. (2014). Pure science and the problem of progress. *Studies in the History and Philosophy of Science* 46:55–63.

Firestein, S. (2012). *Ignorance: How It Drives Science*. Oxford University Press.

Firestein, S. (2015). *Failure: Why Science Is So Successful*. Oxford University Press.

Laudan, L. (1978). *Progress and It's Problems: Towards a Theory of Scientific Growth*. University of California Press.

Levenstein, D., Alvarez, V. A., Amarasingham, A., Azab, H., Gerkin, R. C., Hasenstaub, A., Iyer, R., Jolivet, R. B., Marzen, S., Monaco, J. D., Prinz, A. A., Quraishi, S., Santamaria, F., Shivkumar, S., Singh, M. F., Stockton, D. B., Traub, R., Rotstein, H. G., Nadim, F., and Redish, A. D. (2020). On the role of theory and modeling in neuroscience. arXiv:200313825.

OTHER CITATIONS

Abdul-Hamid, W. K., and Hughes, J. H. (2014). Nothing new under the sun: Post-traumatic stress disorders in the ancient world. *Early Science and Medicine* 19:1–9.

Aimone, J. A., Houser, D., and Weber, B. (2014). Neural signatures of betrayal aversion: An fMRI study of trust. *Proceedings of the Royal Society B: Biological Sciences* 281(1782):20132127.

Akerlof, G. A. (1970). The market for "lemons": Quality uncertainty and the market mechanism. *Quarterly Journal of Economics* 84(3):488–500.

Akerlof, G. A., and Kranton, R. E. (2010). *Identity Economics*. Princeton University Press.

Akerlof, G. A., and Shiller, R. J. (2015). *Phishing for Phools*. Princeton University Press.

Allen, R. C. (1982). The efficiency and distributional consequences of eighteenth century enclosures. *Economic Journal* 92(368):937–953.

Altman, J., Everitt, B. J., Robbins, T. W., Glautier, S., Markou, A., Nutt, D., Oretti, R., and Phillips, G. D. (1996). The biological, social and clinical bases of drug addiction: Commentary and debate. *Psychopharmacology* 125(4):285–345.

Anderson, L. R., and Mellor, J. M. (2009). Religion and cooperation in a public goods experiment. *Economics Letters* 105(1):58–60.

Anticevic, A., and Murray, J., eds. (2017). *Computational Psychiatry*. Elsevier.

Ariely, D. (2008). *Predictably Irrational*. Harper.

Aronson, E., and Mills, J. (1959). The effect of severity of initiation on liking for a group. *Journal of Abnormal and Social Psychology* 59(2):177.

Asch, S. E. (1948). The doctrine of suggestion, prestige and imitation in social psychology. *Psychological Review* 55(5):250.

Asch, S. E. (1952). *Social Psychology*. Prentice Hall.

Augustine of Hippo (Saint Augustine). (398/1961). *Confessions*. Translated by R. S. Pine-Coffin. Penguin Classics.

Augustine of Hippo (Saint Augustine). (427/1972). *The City of God*. Translated by H. Bettenson. Penguin Classics.

Balfour, D. J. K., and Fagerström, K. O. (1996). Pharmacology of nicotine and its therapeutic use in smoking cessation and neurodegenerative disorders. *Pharmacology and Therapeutics* 72(1):51–81.

Bandura, A., Underwood, B., and Fromson, M. E. (1975). Disinhibition of aggression through diffusion of responsibility and dehumanization of victims. *Journal of Research in Personality* 9(4):253–269.

Bargh, J. A., and Chartrand, T. L. (2014). The mind in the middle: A practical guide to priming and automaticity research. In: *Handbook of Research Methods in Social and Personality Psychology*, edited by H. T. Reis and C. M. Judd, 311–344. Cambridge University Press.

Barreca, A. I., Neidell, M., and Sanders, N. J. (2017). Long-run pollution exposure and adult mortality: Evidence from the acid rain program. National Bureau of Economic Research.

Barry, C. L., McGinty, E. E., Pescosolido, B. A., and Goldman, H. H. (2014). Stigma, discrimination, treatment effectiveness, and policy: Public views about drug addiction and mental illness. *Psychiatric Services* 65(10):1269–1272.

Bastiaansen, J. A. C. J., Thioux, M., and Keysers, C. (2009). Evidence for mirror systems in emotions. *Philosophical Transactions of the Royal Society B* 364(1528):2391–2404.

Bateson, J. (2015). *The Last and Greatest Battle: Finding the Will, Commitment, and Strategy to End Military Suicides*. Oxford University Press.

Bateson, M., Nettle, D., and Roberts, G. (2006). Cues of being watched enhance cooperation in a real-world setting. *Biology Letters* 2(3):412–414.

Batson, C. D. (2010). Empathy-induced altruistic motivation. In: *Prosocial Motives, Emotions and Behaviors: The Better Angels of Our Nature*, edited by M. Mikulincer and P. R. Shaver, 15–34. American Psychological Association.

Batson, C. D., Duncan, B. D., Ackerman, P., Buckley, T., and Birch, K. (1981). Is empathic emotion a source of altruistic motivation? *Journal of Personality and Social Psychology* 40(2):290.

Batts, S. (2009). Brain lesions and their implications in criminal responsibility. *Behavioral Sciences and the Law* 27(2):261–272.

Bauer, S. (2018). *American Prison*. Penguin.

Bauman, C. W., McGraw, A. P., Bartels, D. M., and Warren, C. (2014). Revisiting external validity: Concerns about trolley problems and other sacrificial dilemmas in moral psychology. *Social and Personality Psychology Compass* 8(9):536–554.

Baumeister, R. F., Heatherton, T. F., and Tice, D. M. (1994). *Losing Control: How and Why People Fail at Self-Regulation*. Academic Press.

Baumeister, R. F., Vohs, K. D., and Tice, D. M. (2007). The strength model of self-control. *Psychological Science* 16(6):351–355.

Bechara, A. (2005). Decision making, impulse control and loss of willpower to resist drugs: A neurocognitive perspective. *Nature Neuroscience* 8(11):1458–1463.

Beller, J., and Wagner, A. (2018). Loneliness, social isolation, their synergistic interaction, and mortality. *Health Psychology* 37(9):808.

Benowitz, N. L. (1996). Pharmacology of nicotine: Addiction and therapeutics. *Annual Review of Pharmacology and Toxicology* 36:597–613.

Bentham, J. (1789/1907). *An Introduction to the Principles of Morals and Legislation.* Oxford University Press.

Berg, J., Dickhaut, J., and McCabe, K. (1995). Trust, reciprocity, and social history. *Games and Economic Behavior* 10:122–142.

Bergström, A., Jayatissa, M., Mørk, A., and Wiborg, O. (2008). Stress sensitivity and resilience in the chronic mild stress rat model of depression: An in situ hybridization study. *Brain Research* 1196:41–52.

Berman, J. Z., and Small, D. A. (2018). Discipline and desire: On the relative importance of willpower and purity in signaling virtue. *Journal of Experimental Social Psychology* 76:220–230.

Bernard, V., Ottenberg, P., and Redl, F. (1965). Dehumanization: A composite psychological defense in relation to modern war. In: *Behavioral Science and Human Survival,* edited by M. Schwebel, 64–82. Science and Behavior Books.

Bickel, W. K., and Marsch, L. A. (2001). Toward a behavioral economic understanding of drug dependence: Delay discounting processes. *Addiction* 96:73–86.

Bland, A. R., Roiser, J. P., Mehta, M. A., Schei, T., Sahakian, B. J., Robbins, T. W., and Elliott, R. (2017). Cooperative behavior in the ultimatum game and prisoner's dilemma depends on players' contributions. *Frontiers in Psychology* 8:1017.

Blount, S. (1995). When social outcomes aren't fair: The effect of causal attributions on preferences. *Organizational Behavior and Human Decision Processes* 63(2):131–144.

Blum, D. (1995). *The Monkey Wars.* Oxford University Press.

Blum, D. (2002). *Love at Goon Park: Harry Harlow and the Science of Affection.* Perseus Books.

Bolton, G. E., Feldhaus, C., and Ockenfels, A. (2016). Social interaction promotes risk taking in a stag hunt game. *German Economic Review* 17(3):409–423.

Bonanno, G. A., Wortman, C. B., Lehman, D. R., Tweed, R. G., Haring, M., Sonnega, J., Carr, D., and Nesse, R. M. (2002). Resilience to loss and chronic grief: A prospective study from preloss to 18-months postloss. *Journal of Personality and Social Psychology* 83(5):1150.

Bostrom, N. (2014). *Superintelligence: Paths, Dangers, Strategies.* Oxford University Press.

Bosworth, S. J., Singer, T., and Snower, D. J. (2016). Cooperation, motivation, and social balance. *Journal of Economic Behavior and Organization* 126:72–94.

Bouton, M. E. (2007). *Learning and Behavior: A Contemporary Synthesis.* Sinauer.

Boyd, R., Gintis, H., Bowles, S., and Richerson, P. J. (2003). The evolution of altruistic punishment. *Proceedings of the National Academy of Sciences* 100(6):3531–3535.

Braithwaite, J. (2004). Restorative justice and de-professionalization. *Good Society* 13(1):28–31.

Bramble, D. M., and Lieberman, D. E. (2004). Endurance running and the evolution of homo. *Nature* 432(7015):345–352.

Bray, S., Rangel, A., Shimojo, S., Balleine, B., and O'Doherty, J. P. (2008). The neural mechanisms underlying the influence of Pavlovian cues on human decision making. *Journal of Neuroscience* 28(22):5861–5866.

Bruers, S., and Braeckman, J. (2014). A review and systematization of the trolley problem. *Philosophia* 42(2):251–269.

Bruner, N., and Johnson, M. (2013). Demand curves for hypothetical cocaine in cocaine-dependent individuals. *Psychopharmacology* 231(5):889–897.

Bucciol, A., and Piovesan, M. (2011). Luck or cheating? A field experiment on honesty with children. *Journal of Economic Psychology* 32(1):73–78.

Buckner, R. L., and Carroll, D. C. (2007). Self-projection and the brain. *Trends in Cognitive Sciences* 11(2):49–57.

Burger, J., Ostrom, E., Norgaard, R. B., Policansky, D., and Goldstein, B. D., eds. (2001). *Protecting the Commons: A Framework for Resource Management in the Americas*. Island Press.

Burger, J. M. (2009). Replicating Milgram: Would people still obey today? *American Psychologist* 64(1):1.

Burger, J. M., Girgis, Z. M., and Manning, C. C. (2011). In their own words: Explaining obedience to authority through an examination of participants' comments. *Social Psychological and Personality Science* 2(5):460–466.

Burns, J. M., and Swerdlow, R. H. (2003). Right orbitofrontal tumor with pedophilia symptom and constructional apraxia sign. *Archives of Neurology* 60(3):437–440.

Burton-Chellew, M. N., and West, S. A. (2012). Correlates of cooperation in a one-shot high-stakes televised prisoners' dilemma. *PloS ONE* 7(4):e33344.

Cacioppo, J. T., Cacioppo, S., Capitanio, J. P., and Cole, S. W. (2015). The neuroendocrinology of social isolation. *Annual Review of Psychology* 66:733–767.

Cacioppo, S., Capitanio, J. P., and Cacioppo, J. T. (2014). Toward a neurology of loneliness. *Psychological Bulletin* 140(6):1464.

Caldwell, M. (2017). Whispering to psychopaths: Implications of the Mendota model for the treatment of youth with callous/unemotional symptoms. Working paper, 36th Annual Research and Treatment Conference of the Association for the Treatment of Sexual Abusers. www.wiatsa .org/wp-content/uploads/2017/10/T-30-Whispering-to-Psychopaths.pdf.

Call, J., Carpenter, M., and Tomasello, M. (2005). Copying results and copying actions in the process of social learning: Chimpanzees (Pan troglodytes) and human children (Homo sapiens). *Animal Cognition* 8(3):151–163.

Campbell, J. (1959–1968). *The Masks of God*. 4 vols. Penguin.

Campo, J. L. (2015). From a distance: The psychology of killing with remotely piloted aircraft. PhD diss., Air University.

Capraro, V., Jordan, J. J., and Rand, D. G. (2014). Heuristics guide the implementation of social preferences in one-shot prisoner's dilemma experiments. *Scientific Reports* 4:6790.

Caro, R. (2013). *The Passage of Power*. Vol. 4 of *The Years of Lyndon Johnson*. Vintage Books.

Carruthers, S. L. (2005). Between camps: Eastern bloc "escapees" and cold war borderlands. *American Quarterly* 57(3):911–942.

Carter, G. G., Farine, D. R., Crisp, R. J., Vrtilek, J. K., Ripperger, S. P., and Page, R. A. (2020). Development of new food-sharing relationships in vampire bats. *Current Biology* 30(6):1275–1279.e3.

Carter, G. G., and Wilkinson, G. (2013). Does food sharing in vampire bats demonstrate reciprocity? *Communicative and Integrative Biology* 6(6):e25783.

Chang, S. W., Gariépy, J. F., and Platt, M. L. (2013). Neuronal reference frames for social decisions in primate frontal cortex. *Nature Neuroscience* 16(2):243–250.

Chappelle, W., Goodman, T., Reardon, L., and Thompson, W. (2014). An analysis of post-traumatic stress symptoms in United States Air Force drone operators. *Journal of Anxiety Disorders* 28(5):480–487.

Charney, D. S. (2004). Psychobiological mechanisms of resilience and vulnerability: Implications for successful adaptation to extreme stress. *American Journal of Psychiatry* 161(2):195–216.

Chartrand, T. L., and Bargh, J. A. (1999). The chameleon effect: The perception-behavior link and social interaction. *Journal of Personality and Social Psychology* 76(6):893–910.

Classen, C., Koopman, C., and Spiegel, D. (1993). Trauma and dissociation. *Bulletin of the Menninger Clinic* 57(2):178.

Coates, J. (2012). *The Hour between Dog and Wolf*. Penguin.

Coates, T. N. (2014). The case for reparations. *Atlantic*.

Coates, T. N. (2015). *Between the World and Me*. Speigel and Grau.

Contopoulos-Ioannidis, D. G., Alexiou, G. A., Gouvias, T. C., and Ioannidis, J. (2008). Life cycle of translational research for medical interventions. *Science* 321(5894):1298–1299.

Cooper, J. C., Kreps, T. A., Wiebe, T., Pirkl, T., and Knutson, B. (2010). When giving is good: Ventromedial prefrontal cortex activation for others' intentions. *Neuron* 67(3):511–521.

Coppola, F. (2019). The brain in solitude: An (other) Eighth Amendment challenge to solitary confinement. *Journal of Law and the Biosciences* 6(1):184–225.

Costa, A., Foucart, A., Hayakawa, S., Aparici, M., Apesteguia, J., Heafner, J., and Keysar, B. (2014). Your morals depend on language. *PLoS ONE* 9(4):e94842.

Crockett, M. J., Braams, B. R., Clark, L., Tobler, P. N., Robbins, T. W., and Kalenscher, T. (2013). Restricting temptations: Neural mechanisms of precommitment. *Neuron* 79(2):391–401.

Crocq, M. A., and Crocq, L. (2000). From shell shock and war neurosis to posttraumatic stress disorder: A history of psychotraumatology. *Dialogues in Clinical Neuroscience* 2(1):47.

Curry, O. S. (2016). Morality as cooperation: A problem-centred approach. In: *The Evolution of Morality*, 27–51. Springer.

Da Costa, J. M. (1884). *Medical Diagnosis, with Special Reference to Practical Medicine*. Lippincott.

Davidai, S., Gilovich, T., and Ross, L. D. (2012). The meaning of default options for potential organ donors. *Proceedings of the National Academy of Sciences* 109(38):15201–15205.

Davis, M., Falls, W. A., Campeau, S., and Kim, M. (1993). Fear-potentiated startle: A neural and pharmacological analysis. *Behavioural Brain Research* 58(1–2):175–198.

De Dreu, C. K., Greer, L. L., Van Kleef, G. A., Shalvi, S., and Handgraaf, M. J. (2011). Oxytocin promotes human ethnocentrism. *Proceedings of the National Academy of Sciences* 108(4):1262–1266.

Delehanty, C., Mewhirter, J., Welch, R., and Wilks, J. (2017). Militarization and police violence: The case of the 1033 program. *Research and Politics* 4(2):1–7.

DePaulo, B. M. (1992). Nonverbal behavior and self-presentation. *Psychological Bulletin* 111(2):203.

Desimone, R. (1996). Neural mechanisms for visual memory and their role in attention. *Proceedings of the National Academy of Sciences* 93(24):13494–13499.

Dixon, T. (2008). *Science and Religion: A Very Short Introduction*. Oxford University Press.

Doll, B. B., Duncan, K. D., Simon, D. A., Shohamy, D., and Daw, N. D. (2015). Model-based choices involve prospective neural activity. *Nature Neuroscience* 18:767–772.

Domjan, M. (1998). *The Principles of Learning and Behavior*. 4th ed. Brooks/Cole.

Dunbar, R. (1996). *Grooming, Gossip, and the Evolution of Language*. Harvard University Press.

Dunmore, E., Clark, D. M., and Ehlers, A. (1999). Cognitive factors involved in the onset and maintenance of posttraumatic stress disorder (PTSD) after physical or sexual assault. *Behaviour Research and Therapy* 37(9):809–829.

Durant, W. (1930). *The Case for India*. Simon and Schuster.

Edelman, G. M. (1992). *Bright Air, Brilliant Fire: On the Matter of the Mind*. Basic Books.

Eisenberger, N. I., Lieberman, M. D., and Williams, K. D. (2003). Does rejection hurt? An fMRI study of social exclusion. *Science* 302(5643):290–292.

Elmontaite, I., Schindler, I., Marco, M. D., Duzzi, D., Venneri, A., and Goel, V. (2019). Left amygdala and putamen activation modulate emotion driven decisions in the iterated prisoner's dilemma game. *Frontiers in Neuroscience* 13:741.

Emler, N. (1994). Gossip, reputation, and social adaptation. In: *Good Gossip*, edited by R. F. Goodman and A. Ben-Ze'ev, 117–138. University of Kansas Press.

Ericsson, K. A., ed. (2006). *The Cambridge Handbook of Expertise and Expert Performance*. Cambridge University Press.

Ericsson, K. A., Krampe, R., and Tesch-Römer, C. (1993). The role of deliberate practice in the acquisition of expert performance. *Psychological Review* 100(3):363–406.

Evans, S. L., Dal Monte, O., Noble, P., and Averbeck, B. B. (2014). Intranasal oxytocin effects on social cognition: A critique. *Brain Research* 1580:69–77.

Fareri, D. S., Chang, L. J., and Delgado, M. R. (2012). Effects of direct social experience on trust decisions and neural reward circuitry. *Frontiers in Neuroscience* 6:148.

Feder, A., Nestler, E. J., Westphal, M., and Charney, D. S. (2010). Psychobiological mechanisms of resilience to stress. In: *Handbook of Adult Resilience*, edited by J. W. Reich, A. J. Zautra, and J. S. Hall, 35–54. Guildford Press.

Feeny, N. C., Zoellner, L. A., and Foa, E. B. (2000). Anger, dissociation, and posttraumatic stress disorder among female assault victims. *Journal of Traumatic Stress* 13(1):89–100.

Fehr, E., and Camerer, C. F. (2007). Social neuroeconomics: The neural circuitry of social preferences. *Trends in Cognitive Sciences* 11(10):419–427c.

Feldman, R. (2012). Oxytocin and social affiliation in humans. *Hormones and Behavior* 61(3):380–391.

Fernandes, S., Aharoni, E., Harenski, C. L., Caldwell, M., and Kiehl, K. A. (2020). Anomalous moral intuitions in juvenile offenders with psychopathic traits. *Journal of Research in Personality* 86:103962.

Festinger, L. (1957). *A Theory of Cognitive Dissonance*. Stanford University Press.

Fine, C. (2010). *Delusions of Gender*. W. W. Norton.

Fischbacher, U., and Föllmi-Heusi, F. (2013). Lies in disguise—an experimental study on cheating. *Journal of the European Economic Association* 11(3):525–547.

Fischbacher, U., Gächter, S., and Fehr, E. (2001). Are people conditionally cooperative? Evidence from a public goods experiment. *Economics Letters* 71(3):397–404.

Fischer, P., Krueger, J. I., Greitemeyer, T., Vogrincic, C., Kastenmüller, A., Frey, D., Heene, M., Wicher, M., and Kainbacher, M. (2011). The bystander-effect: A meta-analytic review on bystander intervention in dangerous and non-dangerous emergencies. *Psychological Bulletin* 137(4):517.

Fiske, A. P., and Rai, T. S. (2015). *Virtuous Violence*. Cambridge University Press.

Foner, E. (2010). *The Fiery Trial*. W. W. Norton.

Fontana, A., Rosenheck, R., and Brett, E. (1992). War zone traumas and posttraumatic stress disorder symptomatology. *Journal of Nervous and Mental Disease* 180(12):748–755.

Fontinelle, A. (April 3, 2020). Tax withholding: Good for government, bad for taxpayers. Technical report. *Investopedia*.

Foot, P. (1967). The problem of abortion and the doctrine of double effect. *Oxford Review* 5:5–15.

Forsythe, R., Horowitz, J. L., Savin, N. E., and Sefton, M. (1994). Fairness in simple bargaining experiments. *Games and Economic Behavior* 6(3):347–369.

Foster, E. K. (2004). Research on gossip: Taxonomy, methods, and future directions. *Review of General Psychology* 8(2):78–99.

Fox, J. (2009). *The Myth of the Rational Market.* HarperCollins.

Friedman, B. (2021). Disaggregating the police function. *University of Pennsylvania Law Review* 169(4):925–999.

Friedman, M. (1970). A Friedman doctrine—the social responsibility of business is to increase its profits. *New York Time Magazine*, September 13, 1970.

Frontline. (2009). "Soldier's Heart." PBS.

Gallagher, W. (2017). *How the Post Office Created America.* Penguin.

Gavrielides, T., ed. (2005). *Restorative Justice.* Ashgate.

Gerard, H. B., and Mathewson, G. C. (1966). The effects of severity of initiation on liking for a group: A replication. *Journal of Experimental Social Psychology* 2(3):278–287.

Gęsiarz, F., and Crockett, M. J. (2015). Goal-directed, habitual, and Pavlovian prosocial behavior. *Frontiers in Behavioral Neuroscience* 9:135. 10.3389/fnbeh.2015.00135.

Ghiglieri, M. P. (1988). *East of the Mountains of the Moon.* Free Press.

Ghiselin, M. T. (1974). *The Economy of Nature and the Evolution of Sex.* University of California Press.

Gilbert, D. T., and Wilson, T. D. (2007). Prospection: Experiencing the future. *Science* 317(5843):1351–1354.

Gilbertson, M. W., McFarlane, A. C., Weathers, F. W., Keane, T. M., Yehuda, R., Shalev, A. Y., Lasko, N. B., Goetz, J. M., and Pitman, R. K. (2010). Is trauma a causal agent of psychopathologic symptoms in posttraumatic stress disorder? Findings from identical twins discordant for combat exposure. *Journal of Clinical Psychiatry* 71(10):1324–1330.

Gilbertson, M. W., Shenton, M. E., Ciszewski, A., Kasai, K., Lasko, N. B., Orr, S. P., and Pitman, R. K. (2002). Smaller hippocampal volume predicts pathologic vulnerability to psychological trauma. *Nature Neuroscience* 5:1242–1247.

Gissurarson, H. H. (2000). *Overfishing: The Icelandic Solution.* Institute for Economic Affairs.

Glenn, D. M., Beckham, J. C., Feldman, M. E., Kirby, A. C., Hertzberg, M. A., and Moore, S. D. (2002). Violence and hostility among families of Vietnam veterans with combat-related post-traumatic stress disorder. *Violence and Victims* 17(4):473–489.

Gneezy, U., and Rustichini, A. (2000). A fine is a price. *Journal of Legal Studies* 29:1–18.

Godfrey-Smith, P. (2016). *Other Minds: The Octopus, the Sea, and the Deep Origins of Consciousness.* Farrar, Straus and Giroux.

Gold, J. I., and Shadlen, M. N. (2001). Neural computations that underlie decisions about sensory stimuli. *Trends in Cognitive Sciences* 5(1):10–16.

Goldstein, A. (2000). *Addiction: From Biology to Drug Policy.* Oxford University Press.

Goodall, J. (1986). *The Chimpanzees of Gombe: Patterns of Behavior*. Belknap.

Goodfellow, I., Bengio, Y., and Courville, A. (2016). *Deep Learning*. MIT Press.

Gothard, K. M., Mosher, C. P., Zimmerman, P. E., Putnam, P. T., Morrow, J. K., and Fuglevand, A. J. (2017). New perspectives on the neurophysiology or primate amygdala emerging from the study of naturalistic social behaviors. *Wiley Interdisciplinary Reviews (WIREs) Cognitive Science* 9(1):e1449.

Gottier, R. F. (1968). The dominance-submission hierarchy in the social behavior of the domestic chicken. *Journal of Genetic Psychology* 112(2):205–226.

Gould, S. J. (1981). *The Mismeasure of Man*. W. W. Norton.

Gould, S. J. (1983). *Hen's Teeth and Horse's Toes*. W. W. Norton.

Gould, S. J. (2000). *Rocks of Ages: Science and Religion in the Fullness of Life*. Random House.

Gould, S. J. (2002). *The Structure of Evolutionary Theory*. Belknap.

Graeber, D. (2014). *Debt*. Melville House.

Greene, J. D., Sommerville, R. B., Nystrom, L. E., Darley, J. M., and Cohen, J. D. (2001). An fMRI investigation of emotional engagement and moral judgement. *Science* 293(5537):2105–2108.

Groopman, J. (2007). *How Doctors Think*. Houghton Mifflin.

Grossman, D. A. (1995). *On Killing: The Psychological Cost of Learning to Kill in War and Society*. Little, Brown.

Grossman, M., and Chaloupka, F. J. (1998). The demand for cocaine by young adults: A rational addiction approach. *Journal of Health Economics* 17:427–474.

Grossman, P. J., and Parrett, M. B. (2011). Religion and prosocial behaviour: A field test. *Applied Economics Letters* 18(6):523–526.

Güth, W., Schmittberger, R., and Schwarze, B. (1982). An experimental analysis of ultimatum bargaining. *Journal of Economic Behavior and Organization* 3(4):367–388.

Hagan, B. O., Wang, E. A., Aminawung, J. A., Albizu-Garcia, C. E., Zaller, N., Nyamu, S., Shavit, S., Deluca, J., Fox, A. D., et al. (2018). History of solitary confinement is associated with post-traumatic stress disorder symptoms among individuals recently released from prison. *Journal of Urban Health* 95(2):141–148.

Hamilton, L. C., Haedrich, R. L., and Duncan, C. M. (2004). Above and below the water: Social/ecological transformation in northwest Newfoundland. *Population and Environment* 25(3):195–215.

Hamilton, W. D. (1964). The genetical evolution of social behavior. *Journal of Theoretical Biology* 7:17–52.

Hanes, W. T., and Sanello, F. (2002). *Opium Wars: The Addiction of One Empire and the Humiliation of Another*. Sourcebooks.

Hanson, K., Allen, S., Jensen, S., and Hatsukami, D. (2003). Treatment of adolescent smokers with the nicotine patch. *Nicotine and Tobacco Research* 5(4):515–526.

Harari, Y. N. (2015). *Sapiens: A Brief History of Humankind*. Harper Perennial.

Harbaugh, W. T., Mayr, U., and Burghart, D. R. (2007). Neural responses to taxation and voluntary giving reveal motives for charitable donations. *Science* 316(5831):1622–1625.

Hardin, G. (1968). The tragedy of the commons. *Science* 162(3859):1243–1248.

Harding, G. (1986). Constructing addiction as a moral failing. *Sociology of Health and Illness* 8(1):75–85.

Hare, T. A., Camerer, C. F., Knoepfle, D. T., O'Doherty, J. P., and Rangel, A. (2010). Value computations in ventral medial prefrontal cortex during charitable decision making incorporate input from regions involved in social cognition. *Journal of Neuroscience* 30(2):583–590.

Hare, T. A., Camerer, C. F., and Rangel, A. (2009). Self-control in decision-making involves modulation of the vmPFC valuation system. *Science* 324(5927):646–648.

Harlow, H. F. (1986). *From Learning to Love*. Praeger.

Hart, C. L. (2013). *High Price*. Harper Perennial.

Hart, C. L. (2017). Viewing addiction as a brain disease promotes social injustice. *Nature Human Behaviour* 1(3):1–1.

Hashemi, M. M., Gladwin, T. E., de Valk, N. M., Zhang, W., Kaldewaij, R., van Ast, V., Koch, S. B., Klumpers, F., and Roelofs, K. (2019). Neural dynamics of shooting decisions and the switch from freeze to fight. *Scientific Reports* 9(1):1–10.

Haslam, S. A., and Reicher, S. (2007). Beyond the banality of evil: Three dynamics of an interactionist social psychology of tyranny. *Personality and Social Psychology Bulletin* 33(5):615–622.

Haynos, A. F., Widge, A. S., Anderson, L. M., and Redish, A. D. (2022). Beyond description and deficits: How computational psychiatry can enhance an understanding of decision-making in anorexia nervosa. *Current Psychiatry Reports*. https://doi.org/10.1007/s11920-022-01320-9.

Hebb, D. O. (1949/2002). *The Organization of Behavior*. Wiley/Lawrence Erlbaum.

Heckhausen, J. (2000). Evolutionary perspectives on human motivation. *American Behavioral Scientist* 43(6):1015–1029.

Heim, C., Shugart, M., Craighead, W. E., and Nemeroff, C. B. (2010). Neurobiological and psychiatric consequences of child abuse and neglect. *Developmental Psychobiology* 52(7):671–690.

Henrich, J., Heine, S. J., and Norenzayan, A. (2010). The weirdest people in the world? *Behavioral and Brain Sciences* 33(2–3):61–83.

Henrich, J., and Muthukrishna, M. (2021). The origins and psychology of human cooperation. *Annual Reviews of Psychology* 72:207–240.

Henrich, J., and Silk, J. B. (2013). Interpretative problems with chimpanzee ultimatum game. *Proceedings of the National Academy of Sciences* 110(33):E3049–E3049.

Herodotus. (430 BCE/1996). *Histories*. Translated by J. M. Marincola and A. De Selincourt. Penguin.

Hertz, J., Krogh, A., and Palmer, R. G. (1991). *Introduction to the Theory of Neural Computation*. Addison-Wesley.

Heuer, L., and Orland, A. (2019). Cooperation in the prisoner's dilemma: An experimental comparison between pure and mixed strategies. *Royal Society Open Science* 6(7):182142.

Heyman, G. (2009). *Addiction: A Disorder of Choice*. Harvard University Press.

Higgins, S. T., Delaney, D. D., Budney, A. J., Bickel, W. K., Hughes, J. R., Foerg, F., and Fenwick, J. W. (1991). A behavioral approach to achieving initial cocaine abstinence. *American Journal of Psychiatry* 148:1218–1224.

Hodges, B. H., and Geyer, A. L. (2006). A noncomformist account of the Asch experiments: Values, pragmatics, and moral dilemmas. *Personality and Social Psychology Review* 10(1):2–19.

Hoffman, E., McCabe, K., Shachat, K., and Smith, V. (1994). Preferences, property rights, and anonymity in bargaining games. *Games and Economic Behavior* 7:346–380.

Hoge, E. A., Austin, E. D., and Pollack, M. H. (2007). Resilience: Research evidence and conceptual considerations for posttraumatic stress disorder. *Depression and Anxiety* 24(2):139–152.

Hölldobler, B., and Wilson, E. O. (1990). *The Ants*. Belknap.

Horn, S. R., and Feder, A. (2018). Understanding resilience and preventing and treating PTSD. *Harvard Review of Psychology* 26(3):158–174.

Horowitz, J. (2016). "Every societal failure, we put it on the cops to solve": Dallas police chief David Brown. Quartz. qz.com/729378/dallas-police-chief-david-brown-were-asking-cops-to-do-too-much-in-this-country/.

Huettel, S. A., Song, A. W., and McCarthy, G. (2004). *Functional Magnetic Resonance Imaging*. Sinauer.

Hume, D. (1740/2000). *A Treatise on Human Nature*. Oxford University Press.

Hursh, S. R. (1991). Behavioral economics of drug self-administration and drug abuse policy. *Journal of the Experimental Analysis of Behavior* 56(2):377–393.

Iannaccone, L. R. (1992). Sacrifice and stigma: Reducing free-riding in cults, communes, and other collectives. *Journal of Political Economy* 100(2):271–291.

Insel, T. R. (2000). Toward a neurobiology of attachment. *Review of General Psychology* 2:176–185.

Insel, T. R. (2010). The challenge of translation in social neuroscience: A review of oxytocin, vasopressin, and affiliative behavior. *Neuron* 65:768–779.

Insel, T. R., and Young, L. J. (2001). The neurobiology of attachment. *Nature Reviews Neuroscience* 2(2):129–136.

Inzlicht, M., Shenhav, A., and Olivola, C. Y. (2018). The effort paradox: Effort is both costly and valued. *Trends in Cognitive Sciences* 22(4):337–349.

Jackson, J. C., Watts, J., Henry, T. R., List, J. M., Forkel, R., Mucha, P. J., Greenhill, S. J., Gray, R. D., and Lindquist, K. A. (2019). Emotion semantics show both cultural variation and universal structure. *Science* 366(6472):1517–1522.

Janak, P., and Tye, K. (2015). From circuits to behaviour in the amygdala. *Nature* 517:284–292.

Janis, I. L. (1982). *Groupthink Psychological Studies of Policy Decisions and Fiascos*. Houghton Mifflin.

Jensen, K., Call, J., and Tomasello, M. (2007). Chimpanzees are rational maximizers in an ultimatum game. *Science* 318(5847):107–109.

Johnson, A., and Redish, A. D. (2007). Neural ensembles in CA3 transiently encode paths forward of the animal at a decision point. *Journal of Neuroscience* 27(45):12176–12189.

Johnson-Laird, P. N. (2010). Mental models and human reasoning. *Proceedings of the National Academy of Sciences* 107(43):18243–18250.

Jones, B., and Rachlin, H. (2006). Social discounting. *Psychological Science* 17(4):283–286.

Jones, E., and Wessely, S. (2005). *Shell Shock to PTSD: Military Psychiatry from 1900 to the Gulf War*. Taylor and Francis.

Jordan, J. J., Sommers, R., Bloom, P., and Rand, D. G. (2017). Why do we hate hypocrites? Evidence for a theory of false signaling. *Psychological Science* 28(3):356–368.

Jordan, K. C. (2002). Right for the wrong reasons: S. L. A. Marshall and the ratio of fire in Korea. *Journal of Military History* 66:135–162.

Kawai, M. (1965). Newly acquired pre-cultural behavior of the natural troop of Japanese monkeys on Koshima islet. *Primates* 61:1–30.

Kelton, S. (2020). *The Deficit Myth: Modern Monetary Theory and the Birth of the People's Economy*. Public Affairs.

Kentner, A. C., Cryan, J. F., and Brummelte, S. (2019). Resilience priming: Translational models for understanding resiliency and adaptation to early life adversity. *Developmental Psychobiology* 61(3):350–375.

Kessler, R. C., Sonnega, A., Bonnet, E., Hughes, M., and Nelson, C. B. (1995). Posttraumatic stress disorder in the national comorbidity survey. *Archives of General Psychiatry* 52(12):1048–1060.

Keynes, J. M. (1936). *The General Theory of Employment, Interest, and Money*. Palgrave Macmillan.

King-Casas, B., and Chiu, P. H. (2012). Understanding interpersonal function in psychiatric illness through multiplayer economic games. *Biological Psychiatry* 72(2):119–125.

King-Casas, B., Sharp, C., Lomax-Bream, L., Lohrenz, T., Fonagy, P., and Montague, P. R. (2008). The rupture and repair of cooperation in borderline personality disorder. *Science* 321:806–810.

King-Casas, B., Tomlin, D., Anen, C., Camerer, C. F., Quartz, S. R., and Montague, P. R. (2005). Getting to know you: Reputation and trust in a two-person economic exchange. *Science* 308(5718):78–83.

Kishida, K. T., King-Casas, B., and Montague, P. R. (2010). Neuroeconomic approaches to mental disorders. *Neuron* 67(4):543–554.

Koechlin, E., Ody, C., and Kouneiher, F. (2003). The architecture of cognitive control in the human prefrontal cortex. *Science* 302(5648):1181–1185.

Koob, G. F., and Le Moal, M. (2006). *Neurobiology of Addiction*. Elsevier Academic Press.

Koscik, T. R., and Tranel, D. (2011). The human amygdala is necessary for developing and expressing normal interpersonal trust. *Neuropsychologia* 49(4):602–611.

Kosfeld, M., Heinrichs, M., Zak, P. J., Fischbacher, U., and Fehr, E. (2005). Oxytocin increases trust in humans. *Nature* 435:673–676.

Krajbich, I., Adolphs, R., Tranel, D., Denburg, N. L., and Camerer, C. F. (2009). Economic games quantify diminished sense of guilt in patients with damage to the prefrontal cortex. *Journal of Neuroscience* 29(7):2188–2192.

Kreps, D. M. (1990). Corporate culture and economic theory. In: *Perspectives on Positive Political Economy*, edited by J. E. Alt and K. A. Shepsle, chap. 4, 90–110. Cambridge University Press.

Kross, E., Berman, M. G., Mischel, W., Smith, E. E., and Wager, T. D. (2011). Social rejection shares somatosensory representations with physical pain. *Proceedings of the National Academy of Sciences* 108(15):6270–6275.

Kuhn, T. S. (1962). *The Structure of Scientific Revolutions*. University of Chicago Press.

Kurth-Nelson, Z., Bickel, W. K., and Redish, A. D. (2012). A theoretical account of cognitive effects in delay discounting. *European Journal of Neuroscience* 35:1052–1064.

Kurth-Nelson, Z., and Redish, A. D. (2010). A reinforcement learning model of precommitment in decision making. *Frontiers in Behavioral Neuroscience* 4:184.

Kurth-Nelson, Z., and Redish, A. D. (2012). Don't let me do that! Models of precommitment. *Frontiers in Neuroscience* 6:138.

Laidley, T. T. S. (January 1865). Breech-loading musket. Vol. 3 of *US Service Magazine*. Charles R. Richardson.

LeCun, Y., Bengio, Y., and Hinton, G. (2015). Deep learning. *Nature* 521:436–444.

Leimgruber, K. L., Rosati, A. G., and Santos, L. R. (2016). Capuchin monkeys punish those who have more. *Evolution and Human Behavior* 37(3):236–244.

Leshner, A. I. (1997). Addiction is a brain disease, and it matters. *Science* 278(5335):45–47.

Leukefeld, C., Gullotta, T. P., and Gregrich, J., eds. (2011). *Handbook of Evidence-Based Substance Abuse Treatment in Criminal Justice Settings*. Springer.

Levanon, A., England, P., and Allison, P. (2009). Occupational feminization and pay: Assessing causal dynamics using 1950–2000 US Census data. *Social Forces* 88(2):865–891.

Levenson, T. (2015). *The Hunt for Vulcan*. Random House.

Levine, S. (1957). Infantile experience and resistance to physiological stress. *Science* 126(3270):405.

Levi-Strauss, C. (1970). *The Raw and the Cooked*. Jonathan Cape.

Lewis, J. (1998). Forgiving George Wallace. *New York Times*.

Lewis, J. (2012). *Across That Bridge*. Hachette.

Liang, L., and Chaloupka, F. J. (2002). Differential effects of cigarette price on youth smoking intensity. *Nicotine and Tobacco Research* 4:190–114.

Liebenberg, L. (2012). *The Art of Tracking: The Origin of Science*. New Africa Books.

Lissek, S., Kaczkurkin, A. N., Rabin, S., Geraci, M., Pine, D. S., and Grillon, C. (2014). Generalized anxiety disorder is associated with overgeneralization of classically conditioned fear. *Biological Psychiatry* 75(11):909–915.

Lloyd, W. F. (1832). *Two Lectures on the Checks to Population*. Oxford University Press.

Locey, M. L., Safin, V., and Rachlin, H. (2013). Social discounting and the prisoner's dilemma game. *Journal of the Experimental Analysis of Behavior* 99(1):85–97.

Lohse, J., and Waichman, I. (2020). The effects of contemporaneous peer punishment on cooperation with the future. *Nature Communications* 11(1):1–8.

Loreau, M., Naeem, S., Inchausti, P., Bengtsson, J., Grime, J. P., Hector, A., Hooper, D. U., Huston, M. A., Raffaelli, D., Schmid, B., Tilman, D., and Wardle, D. A. (2001). Biodiversity and ecosystem functioning: Current knowledge and future challenges. *Science* 294(5543):804–808.

Lum, K. L., Polansky, J. R., Jackler, R. K., and Glantz, S. A. (2008). Signed, sealed and delivered: "Big tobacco" in Hollywood, 1927–1951. *Tobacco Control* 17(5):313–323.

Lussier, J. P., Heil, S. H., Mongeon, J. A., Badger, G. J., and Higgins, S. T. (2006). A meta-analysis of voucher-based reinforcement therapy for substance use disorders. *Addiction* 101:192–203.

Lyons, D. E., Young, A. G., and Keil, F. C. (2007). The hidden structure of overimitation. *Proceedings of the National Academy of Sciences* 104(50):19751–19756.

MacDonald, K., and MacDonald, T. M. (2010). The peptide that binds: A systematic review of oxytocin and its prosocial effects in humans. *Harvard Review of Psychiatry* 18(1):1–21.

MacManus, D., Rona, R., Dickson, H., Somaini, G., Fear, N., and Wessely, S. (2015). Aggressive and violent behavior among military personnel deployed to Iraq and Afghanistan: Prevalence and link with deployment and combat exposure. *Epidemiologic Reviews* 37(1):196–212.

Maguen, S., Luxton, D. D., Skopp, N. A., Gahm, G. A., Reger, M. A., Metzler, T. J., and Marmar, C. R. (2011). Killing in combat, mental health symptoms, and suicidal ideation in Iraq War veterans. *Journal of Anxiety Disorders* 25(4):563–567.

Mahmoud, N. N., and Rothenberger, D. (2019). From burnout to well-being: A focus on resilience. *Clinics in Colon and Rectal Surgery* 32(6):415–423.

Manski, C. F., Pepper, J., and Petrie, C. (2001). *Informing America's Policy on Illegal Drugs: What We Don't Know Keeps Hurting Us*. National Academies Press.

March, J. G., and Olsen, J. P. (2011). The logic of appropriateness. In: *The Oxford Handbook of Political Science*, edited by R. E. Goodin, 478–497. Oxford University Press.

Maren, S., and Quirk, G. J. (2004). Neuronal signaling of fear memory. *Nature Reviews Neuroscience* 5(11):844–852.

Marshall, S. L. A. (1947). *Men against Fire*. William Morrow.

Marwell, G., and Ames, R. E. (1981). Economists free ride, does anyone else? *Journal of Public Economics* 15:295–310.

Mayr, E. (2008). *What Evolution Is*. Hachette.

McClure, S. M., Laibson, D. I., Loewenstein, G., and Cohen, J. D. (2004). Separate neural systems value immediate and delayed monetary rewards. *Science* 306(5695):503–507.

McElrath, W., and Tuberville, S. (2020). Poisoning our police: How the militarization mindset threatens constitutional rights and public safety. *Project on Government Oversight*.

McFall, M., Fontana, A., Raskind, M., and Rosenheck, R. (1999). Analysis of violent behavior in Vietnam combat veteran psychiatric inpatients with posttraumatic stress disorder. *Journal of Traumatic Stress* 12(3):501–517.

McNeill, W. H. (1995). *Keeping Together in Time: Dance and Drill in Human History*. ACLS Humanities.

Mekemson, C., and Glantz, S. A. (2002). How the tobacco industry built its relationship with Hollywood. *Tobacco Control* 11(suppl. 1):i81–i91.

Merry, S. E. (1984). Rethinking gossip and scandal. In: *Toward a General Theory of Social Control*, 271–302. Elsevier.

Meyer, R., and Mirin, S. (1979). *The Heroin Stimulus*. Plenum.

Milinski, M. (2016). Reputation, a universal currency for human social interactions. *Philosophical Transactions of the Royal Society B: Biological Sciences* 371(1687):20150100.

Miller, A. G. (2009). Reflections on "Replicating Milgram" (Burger, 2009). *American Psychologist* 64(1):20–27.

Mischel, W. (2014). *The Marshmallow Test*. Little, Brown.

Moll, J., Eslinger, P. J., and de Oliveira-Souza, R. (2001). Frontopolar and anterior temporal cortex activation in a moral judgement task. *Archives of Neuropsychiatry* 59(3-B):657–664.

Moll, J., Krueger, F., Zahn, R., Pardini, M., de Oliveira-Souza, R., and Grafman, J. (2006). Human fronto-mesolimbic networks guide decisions about charitable donation. *Proceedings of the National Academy of Sciences* 103(42):15623–15628.

Morelli, M. F. (1983). Milgram's dilemma of obedience. *Metaphilosophy* 14(3/4):183–189.

Müller-Leinß, J. M., Enzi, B., Flasbeck, V., and Brüne, M. (2018). Retaliation or selfishness? An rTMS investigation of the role of the dorsolateral prefrontal cortex in prosocial motives. *Social Neuroscience* 13(6):701–709.

Mummolo, J. (2018). Militarization fails to enhance police safety or reduce crime but may harm police reputation. *Proceedings of the National Academy of Sciences* 115(37):9181–9186.

Myers, K. M., and Davis, M. (2002). Behavioral and neural analysis of extinction. *Neuron* 36(4):567–584.

Neto, M. L., Antunes, M., Lopes, M., Ferrieira, D., Rilling, J., and Prata, D. (2020). Oxytocin and vasopressin modulation of prisoner's dilemma strategies. *Journal of Psychopharmacology* 34(8):891–900.

Neumann, I. D., Maloumby, R., Beiderbeck, D. I., Lukas, M., and Landgraf, R. (2013). Increased brain and plasma oxytocin after nasal and peripheral administration in rats and mice. *Psychoneuroendocrinology* 38(10):1985–1993.

Nielsen, M., Simcock, G., and Jenkins, L. (2008). The effect of social engagement on 24-month-olds' imitation from live and televised models. *Developmental Science* 11(5):722–731.

Nordell, J. (2017). Is this how discrimination ends? *Atlantic.*

Norton, M. I., and Ariely, D. (2011). Building a better America—one wealth quintile at a time. *Perspectives on Psychological Science* 6(9):9–12.

Nutt, D. J. (2000). The psychobiology of posttraumatic stress disorder. *Journal of Clinical Psychiatry* 61(suppl. 5):24–29.

O'Connell, R. L. (2010). *The Ghosts of Cannae.* Random House.

O'Craven, K. M., and Kanwisher, N. (2000). Mental imagery of faces and places activates corresponding stimulus-specific brain regions. *Journal of Cognitive Neuroscience* 12(6):1013–1023.

Oosterbeek, H., Sloof, R., and Van De Kuilen, G. (2004). Cultural differences in ultimatum game experiments: Evidence from a meta-analysis. *Experimental Economics* 7(2):171–188.

Osborne, M. J. (2004). *An Introduction to Game Theory.* Oxford University Press.

Ostrom, E., Garnder, R., and Walker, J. (1994). *Rules, Games, and Common-Pool Resources.* University of Michigan Press.

Ostrom, E., Walker, J., and Gardner, R. (1992). Covenants with and without a sword: Self-governance is possible. *American Political Science Review* 86(2):404–417.

Pados, B. F. (2019). Physiology of stress and use of skin-to-skin care as a stress-reducing intervention in the NICU. *Nursing for Women's Health* 23(1):59–70.

Park, C. L. (2010). Making sense of the meaning literature: An integrative review of meaning making and its effects on adjustment to stressful life events. *Psychological Bulletin* 136(2):257.

Parker, K. J., and Maestripieri, D. (2011). Identifying key features of early stressful experiences that produce stress vulnerability and resilience in primates. *Neuroscience and Biobehavioral Reviews* 35(7):1466–1483.

Pavlov, I. (1927). *Conditioned Reflexes.* Oxford University Press.

Pearson, J., Naselaris, T., Holmes, E. A., and Kosslyn, S. M. (2015). Mental imagery: Functional mechanisms and clinical applications. *Trends in Cognitive Sciences* 19(10):590–602.

Pernu, T. K., and Elzein, N. (2020). From neuroscience to law: Bridging the gap. *Frontiers in Psychology* 11:1862. 10.3389/fpsyg.2020.01862.

Peters, J., and Büchel, C. (2010). Episodic future thinking reduces reward delay discounting through an enhancement of prefrontal-mediotemporal interactions. *Neuron* 66(1):138–148.

Peters, J., and Büchel, C. (2011). The neural mechanisms of inter-temporal decision-making: Understanding variability. *Trends in Cognitive Sciences* 15(5):227–239.

Petry, N. M. (2012). *Contingency Management: For Substance Abuse Treatment*. Routledge.

Philpot, R., Liebst, L. S., Levine, M., Bernasco, W., and Lindegaard, M. R. (2019). Would I be helped? Cross-national CCTV footage shows that intervention is the norm in public conflicts. *American Psychologist* 75(1):66–75.

Pitman, R. K., Gilbertson, M. W., Gurvits, T. V., May, F. S., Lasko, N. B., Metzger, L. J., Shenton, M. E., Yehuda, R., and Orr, S. P. (2006). Clarifying the origin of biological abnormalities in PTSD through the study of identical twins discordant for combat exposure. *Annals of the New York Academy of Sciences* 1071:242–254.

Pizarro, J., Silver, R. C., and Prause, J. (2006). Physical and mental health costs of traumatic war experiences among Civil War veterans. *Archives of General Psychiatry* 63(2):193–200.

Platt, S. R. (2018). *Imperial Twilight: The Opium War and the End of China's Last Golden Age*. Knopf.

Popper, K. R. (1959). *The Logic of Scientific Discovery*. Routledge.

Potter, G. (2013). The history of policing in the United States. EKU Online. https://ekuonline.eku .edu/blog/police-studies/the-history-of-policing-in-the-united-states-part-1/.

Proctor, D., Williamson, R. A., de Waal, F. B., and Brosnan, S. F. (2013). Chimpanzees play the ultimatum game. *Proceedings of the National Academy of Sciences* 110(6):2070–2075.

Queller, D. C., and Strassmann, J. E. (1998). Kin selection and social insects. *BioScience* 48(3):165–175.

Rachlin, H., and Jones, B. A. (2008a). Altruism among relatives and non-relatives. *Behavioural Processes* 79(2):120–123.

Rachlin, H., and Jones, B. A. (2008b). Social discounting and delay discounting. *Journal of Behavioral Decision Making* 21(1):29–43.

Radu, P. T., Yi, R., Bickel, W. K., Gross, J. J., and McClure, S. M. (2011). A mechanism for reducing delay discounting by altering temporal attention. *Journal of the Experimental Analysis of Behavior* 96(3):363–385.

Rand, A. (1957). *Atlas Shrugged*. Random House.

Rand, D. G. (2016). Cooperation, fast and slow: Meta-analytic evidence for a theory of social heuristics and self-interested deliberation. *Psychological Science* 27(9):1192–1206.

Rangel, A., and Hare, T. (2010). Neural computations associated with goal-directed choice. *Current Opinion in Neurobiology* 20(2):262–270.

Rapoport, A., and Chammah, A. M. (1965). *Prisoner's Dilemma*. University of Michigan Press.

Redish, A. D. (2013a). The dangers of dualism: Implications of the multiple decision-making system theory for free will and responsibility. *Cognitive Critique* 7:1–28.

Redish, A. D. (2016). Vicarious trial and error. *Nature Reviews Neuroscience* 17:147–159.

Redish, A. D. (2020). Addiction from a computational perspective. In: *Computational Psychiatry: A Primer*, edited by P. Series, chap. 9, 185–204. MIT Press.

Redish, A. D., and Gordon, J. A., eds. (2016). *Computational Psychiatry: New Perspectives on Mental Illness*. Strüngmann Forum Reports. MIT Press.

Redish, A. D., Jensen, S., and Johnson, A. (2008). A unified framework for addiction: Vulnerabilities in the decision process. *Behavioral and Brain Sciences* 31:415–487.

Redish, A. D., Jensen, S., Johnson, A., and Kurth-Nelson, Z. (2007). Reconciling reinforcement learning models with behavioral extinction and renewal: Implications for addiction, relapse, and problem gambling. *Psychological Review* 114(3):784–805.

Redish, A. D., Kummerfeld, E., Morris, R. L., and Love, A. C. (2018). Opinion: Reproducibility failures are essential to scientific inquiry. *Proceedings of the National Academy of Sciences* 115(20):5042–5046.

Regier, P. S., and Redish, A. D. (2015). Contingency management and deliberative decision-making processes. *Frontiers in Psychiatry* 6:76.

Reicher, S., and Haslam, S. A. (2011). After shock? Towards a social identity explanation of the Milgram 'obedience' studies. *British Journal of Social Psychology* 50:163–169.

Reilly, B., Rickman, N., and Witt, R. (2012). Robbing banks: Crime does pay—but not very much. *Significance* 9(3):17–21.

Riedl, K., Jensen, K., Call, J., and Tomasello, M. (2012). No third-party punishment in chimpanzees. *Proceedings of the National Academy of Sciences* 109(37):14824–14829.

Rilling, J. K. (2013). The neural and hormonal bases of human parental care. *Neuropsychologia* 51(4):731–747.

Rizzolatti, G., and Craighero, L. (2004). The mirror-neuron system. *Annual Review of Neuroscience* 27:169–192.

Robbins, T. W., and Clark, L. (2015). Behavioral addictions. *Current Opinion in Neurobiology* 30:66–72.

Robinson, T. E. (2004). Neuroscience: Addicted rats. *Science* 305(5686):951–953.

Rockenbach, B., and Milinski, M. (2006). The efficient interaction of indirect reciprocity and costly punishment. *Nature* 444(7120):718–723.

Roelofs, K. (2017). Freeze for action: Neurobiological mechanisms in animal and human freezing. *Philosophical Transactions of the Royal Society B: Biological Sciences* 372(1718):20160206.

Roth, A. E., and Schoumaker, F. (1983). Expectations and reputations in bargaining: An experimental study. *American Economic Review* 73(3):362–372.

Rothbard, M. N. (2002). Milton Friedman unraveled. *Journal of Libertarian Studies* 16(4):37–54.

Runciman, W. G., and Sen, A. K. (1965). Games, justice and the general will. *Mind* 74(296):554–562.

Runge, C. F. (1981). Common property externalities: Isolation, assurance, and resource depletion in a traditional grazing context. *American Journal of Agricultural Economics* 63(4):595–606.

Runge, C. F. (1984b). Strategic interdependence in models of property rights. *American Journal of Agricultural Economics* 66(5):807–813.

Runge, C. F., and Defrancesco, E. (2006). Exclusion, inclusion, and enclosure: Historical commons and modern intellectual property. *World Development* 34(10):1713–1727.

Russell, N. J. C. (2011). Milgram's obedience to authority experiments: Origins and early evolution. *British Journal of Social Psychology* 50(1):140–162.

Russo, S. J., Murrough, J. W., Han, M. H., Charney, D. S., and Nestler, E. J. (2012). Neurobiology of resilience. *Nature Neuroscience* 15(11):1475–1484.

Rutledge, I. (2006). *Addicted to Oil: America's Relentless Drive for Energy Security*. St. Martin's Press.

Sanfey, A. G. (2007). Social decision-making: Insights from game theory and neuroscience. *Science* 318(5850):598–602.

Sanfey, A. G., Rilling, J. K., Aronson, J. A., Nystrom, L. E., and Cohen, J. D. (2003). The neural basis of economic decision-making in the ultimatum game. *Science* 300(5626):1755–1758.

Sapolsky, R. (1998). Junk food monkeys. In: *The Trouble with Testosterone*. Touchstone.

Schacter, D. L., Addis, D. R., and Buckner, R. L. (2007). Remembering the past to imagine the future: The prospective brain. *Nature Reviews Neuroscience* 8:657–661.

Schmalensee, R., and Stavins, R. N. (2017). Lessons learned from three decades of experience with cap and trade. *Review of Environmental Economics and Policy* 11(1):59–79.

Schüll, N. D. (2012). *Addiction by Design: Machine Gambling in Las Vegas*. Princeton University Press.

Searle, J. (2006). *Freedom and Neurobiology: Reflections on Free Will, Language, and Political Power*. Columbia University Press.

Series, P., ed. (2020). *Computational Psychiatry: A Primer*. MIT Press.

Seymour, B., Singer, T., and Dolan, R. (2007). The neurobiology of punishment. *Nature Reviews Neuroscience* 8(4):300–311.

Shadlen, M. N., and Shohamy, D. (2016). Decision making and sequential sampling from memory. *Neuron* 90(5):927–939.

Shay, J. (1994). *Achilles in Vietnam*. Scribner.

Shay, J. (2002). *Odysseus in America*. Scribner.

Shiller, R. J. (2017). Narrative economics. *American Economic Review* 107(4):967–1004.

Shiromani, P. J., Keane, T. M., and Ledoux, J. E., eds. (2009). *Post-Traumatic Stress Disorder: Basic Science and Clinical Practice*. Humana.

Simoons, F. J. (1994). *Eat Not This Flesh*. University of Wisconsin Press.

Singer, P. (1975). *Animal Liberation*. HarperCollins.

Singer, P. (1981). *The Expanding Circle*. Farrar, Straus and Giroux. Reprinted in 2011 by Princeton University Press.

Singer, T. (2008). Understanding others: Brain mechanisms of theory of mind and empathy. In: *Neuroeconomics: Decision Making and the Brain*, edited by P. W. Glimcher, C. Camerer, R. A. Poldrack, and E. Fehr, 251–268. Academic Press.

Slobogin, C. (2017). Neuroscience nuance: Dissecting the relevance of neuroscience in adjudicating criminal culpability. *Journal of Law and the Biosciences* 4(3):577–593.

Smith, V. L. (2008). *Rationality in Economics*. Cambridge University Press.

Snider, S. E., LaConte, S. M., and Bickel, W. K. (2016). Episodic future thinking: Expansion of the temporal window in individuals with alcohol dependence. *Alcoholism: Clinical and Experimental Research* 40(7):1558–1566.

Sosis, R., and Bressler, E. R. (2003). Cooperation and commune longevity: A test of the costly signaling theory of religion. *Cross-Cultural Research* 37(2):211–239.

Speitel, C., Traut-Mattausch, E., and Jonas, E. (2019). Functions of the right dlPFC and right TPJ in proposers and responders in the ultimatum game. *Social Cognitive and Affective Neuroscience* 14(3):263–270.

Stein, J. S., Wilson, A. G., Koffarnus, M. N., Daniel, T. O., Epstein, L. H., and Bickel, W. K. (2016). Unstuck in time: Episodic future thinking reduces delay discounting and cigarette smoking. *Psychopharmacology* 233(21–22):3771–3778.

Stewart, N., Chater, N., and Brown, G. D. (2006). Decision by sampling. *Cognitive Psychology* 53(1):1–26.

Stone, I. F. (1989). *The Trial of Socrates*. Doubleday.

Stoughton, S. W. (2016). Principled policing: Warrior cops and guardian officers. *Wake Forest Law Review* 51:611.

Stout, L. A. (2008). Taking conscience seriously. In: *Moral Markets*, edited by P. J. Zak, 157–172. Princeton University Press.

Striedter, G. F. (2005). *Principles of Brain Evolution*. Sinauer.

Suchak, M., Eppley, T. M., Campbell, M. W., Feldman, R. A., Quarles, L. F., and de Waal, F. B. (2016). How chimpanzees cooperate in a competitive world. *Proceedings of the National Academy of Sciences* 113(36):10215–10220.

Talarico, J. M., and Rubin, D. C. (2003). Confidence, not consistency, characterizes flashbulb memories. *Psychological Science* 14(5):455–461.

Terris, E. T., Beavin, L. E., Barraza, J. A., Schloss, J., and Zak, P. J. (2018). Endogenous oxytocin release eliminates in-group bias in monetary transfers with perspective taking. *Frontiers in Behavioral Neuroscience* 12:35. 10.3389/fnbeh.2018.00035.

Tetlock, P. E. (2003). Thinking the unthinkable: Sacred values and taboo cognitions. *Trends in Cognitive Sciences* 7(7):320–324.

Tetlock, P. E., Kristel, O. V., Elson, S. B., Green, M. C., and Lerner, J. S. (2000). The psychology of the unthinkable: Taboo trade-offs, forbidden base rates, and heretical counterfactuals. *Journal of Personality and Social Psychology* 78(5):853.

Thaler, R. H., and Benartzi, S. (2007). The behavioral economics of retirement savings behavior. Technical report #2007-02. AARP Public Policy Institute.

Thucydides. (403 BCE/1954). *History of the Peloponnesian War*. Translated by R. Warner. Penguin.

Tilman, D., Isbell, F., and Cowles, J. M. (2014). Biodiversity and ecosystem functioning. *Annual Review of Ecology, Evolution, and Systematics* 45:471–493.

Tomova, L., Wang, K. L., Thompason, T., Matthews, G. A., Takahashi, A., Tye, K. M., and Saxe, R. (2020). Acute social isolation evokes midbrain craving responses similar to hunger. *Nature Neuroscience* 23:1597–1605.

Trabandt, M., and Uhlig, H. (2011). The Laffer curve revisited. *Journal of Monetary Economics* 58(4):305–327.

Traulsen, A., Röhl, T., and Milinski, M. (2012). An economic experiment reveals that humans prefer pool punishment to maintain the commons. *Proceedings of the Royal Society B: Biological Sciences* 279(1743):3716–3721.

Trope, Y., and Liberman, N. (2003). Temporal construal. *Psychological Review* 110(3):403–421.

Turco, R. P., Toon, O. B., Ackerman, T. P., Pollack, J. B., and Sagan, C. (1983). Nuclear winter: Global consequences of multiple nuclear explosions. *Science* 222(4630):1283–1292.

Tusche, A., Böckler, A., Kanske, P., Trautwein, F. M., and Singer, T. (2016). Decoding the charitable brain: Empathy, perspective taking, and attention shifts differentially predict altruistic giving. *Journal of Neuroscience* 36(17):4719–4732.

Uchibe, E., and Doya, K. (2008). Finding intrinsic rewards by embodied evolution and constrained reinforcement learning. *Neural Networks* 21(10):1447–1455.

Ustinova, Y., and Cardeña, E. C. (2014). Combat stress disorders and their treatment in ancient Greece. *Psychological Trauma: Theory, Research, Practice, and Policy* 6(6):739–748.

van den Bos, W., and Güroğlu, B. (2009). The role of the ventral medial prefrontal cortex in social decision making. *Journal of Neuroscience* 29(24):7631–7632.

van Dijk, E., and De Dreu, C. K. W. (2021). Experimental games and social decision making. *Annual Review of Psychology* 72:415–438.

van Meurs, B., Wiggert, N., Wicker, I., and Lissek, S. (2014). Maladaptive behavioral consequences of conditioned fear-generalization: A pronounced, yet sparsely studied, feature of anxiety pathology. *Behaviour Research and Therapy* 57:29–37.

Vico, G. (1725/1744/1948). *The New Science*. Translated by T. G. Bergin and M. H. Fisch. Cornell University Press.

Volkow, N. (2014). *Drugs, Brains, and Behavior: The Science of Addiction*. NIH-Pub No. 14–5605. National Institute on Drug Abuse.

Volkow, N. D., and Li, T. K. (2004). Drug addiction: The neurobiology of behaviour gone awry. *Nature Reviews Neuroscience* 5:963–970.

von Neumann, J., and Morganstern, O. (1944). *Theory of Games and Economic Behavior*. Princeton University Press.

von Rohr, C. R., Koski, S. E., Burkart, J. M., Caws, C., Fraser, O. N., Ziltener, A., and Van Schaik, C. P. (2012). Impartial third-party interventions in captive chimpanzees: A reflection of community concern. *PloS ONE* 7(3):e32494.

Wang, X. (2009). Retail return policy, endowment effect, and consumption propensity: An experimental study. *Behavioral Economics Journal of Economic Analysis and Policy* 9(1). 10.2202/1935-1682.2288.

Wanniski, J. (1978). Taxes, revenues, and the Laffer curve. *Public Interest* 50:3.

Wapner, J. (2020). *Wall Disease: The Psychological Toll of Living Up against a Border*. The Experiment.

Wiener, J. (1995). *The Beak of the Finch: A Story of Evolution in Our Time*. Vintage Books.

Wiggs, C. L., and Martin, A. (1998). Properties and mechanisms of perceptual priming. *Current Opinion in Neurobiology* 8(2):227–233.

Wilson, E. O. (1980). *Sociobiology*. Harvard University Press.

Worchel, S., and Brehm, J. W. (1970). Effect of threats to attitudinal freedom as a function of agreement with the communicator. *Journal of Personality and Social Psychology* 14(1):18.

Xia, N., and Li, H. (2018). Loneliness, social isolation, and cardiovascular health. *Antioxidants and Redox Signaling* 28(9):837–851.

Yamagishi, T. (1986). The provisioning of a sanctioning system as a public good. *Journal of Personality and Social Psychology* 51(1):110–116.

Yehuda, R., and Ledoux, J. (2007). Response variation following trauma: A translational neuroscience approach to PTSD. *Neuron* 56(1):19–32.

Young, H. P. (2015). The evolution of social norms. *Economics* 7(1):359–387.

Young, L. J., Wang, Z., and Insel, T. R. (1998). Neuroendocrine bases of monogamy. *Trends in Neurosciences* 21(2):71–75.

Zak, P. J., Stanton, A. A., and Ahmadi, S. (2007). Oxytocin increases generosity in humans. *PloS ONE* 2(11):e1128.

Zarnowitz, V. (1992). *Business Cycles: Theory, History, Indicators, and Forecasting*. National Bureau of Economic Research.

Index

Morality. *See also* Science of morality
  based on harm avoidance or utility increas-
    ing, 174
  collectivism and, 176
  first-person self-control and, 87
  Haidt's moral matrix of, 178–179
  as a means to solve the assurance game, 265,
    268, 291
  new conceptual framework of, 8–10
  scientific descriptions of, 177–178
  scientific understanding of, 9
Morality experiments. *See also* Economic
    games; Games; Trolley problem; Trustee
    game; Ultimatum game
  Burger's replication study, 107
  changing the question and, 109–112
  by Stanley Milgram, 103–112
Moral nihilism, 268
Moral Origins (Boehm), 83, 217
Moral philosophy, 5, 6, 7
Moral praise, 69–70
Moral questions. *See also* Morality experiments
  on animal welfare, 254–257
  decision-making systems and, 88, 93
  defined by government, 246–247
  economic games asking, 3–4
  interactions with others and, 17, 181–182
    (*see also* Social structure(s))
  sacrificing for others, 116–118
Moral reasoning, 119–122
Moral relativism, 7, 10, 129, 265–272
Moral structures, 171. *See also* Social
    construct(ions); Social institutions and
    policies
  bringing people into an assurance game, 268
  by the Catholic Church, 210–211
  determining impact of, 293
  impact of, 266
  moral relativism and, 266–267
  some better than others, 266, 267
  as tools to solve the assurance game,
    173–174

Moral technologies. *See* Moral structures
Moral tools, 64, 180–181. *See also*
    Punishment(s); Religion(s); Social control
  to address climate change, 257
  dissent and protest, 82
  Pinker on, 179–180
  whistleblowers, 82
Motivation, moral praise/blame and, 69–70
Multiplayer games, 169. *See also* Public goods
    game; Tax game (public goods game)
Multiple decision systems, 3, 87–88, 91,
    93–98, 200, 312
Multiple-decision systems theory, 100, 200,
    201
Murder, police violence and, 120–122

National Basketball Association (NBA), 277
National Football League (NFL), 277
Naturalistic fallacy, 128–129, 173, 177, 178
Nazi Germany/Nazi death camp guards, 11,
    103–105, 104, 109, 246, 275, 288
Nelson, Craig, 272
Neuroeconomics, 3–5, 4–5, 183
Neurophysiology
  deliberation and, 94–95
  empathy and, 146, 150, 152, 252
  instinctual (Pavlovian) and, 96–97
  larger cooperative groups and, 81
  overview, 170
  oxytocin and, 152–153, 251–252
  pattern completion processes, 98
  procedural learning and, 95–96, 211–212
  rules and, 98
  sharing within groups and, 80
  trolley problem, 112–114
  of trust, 148–153
Neuroscience/psychology, 7, 13. *See also*
    Brain; Decision-making systems; Mam-
    malian brain
  decision-making systems and, 10, 93–98,
    101
  dual-system theories, 98–101

Sports games, 19, 20
  group competition, 277–278
  playing with asabiya, 135
  procedural decision-making systems and, 95, 96
  teams and, 56
Stag hunt game, 29–30
Star Trek: Deep Space Nine (TV series), 104, 310–311
Star Trek: The Next Generation (TV series), 261
Status, 136–139
Stephan, Maria, 83
Stories, empathy and, 147–148
Stout, Lynn, 52, 59
Strangers, 80–81, 149, 150, 252–253
Subgroup(s), 41, 191, 237, 243, 279
Success, intrinsic goal of, 136–137
Suckers
  in prisoner's dilemma game, 24, 295
  tax game and, 37–38
Suicide, 156, 158, 160
Super-altruists, 149–150
Superintelligence (Bostrom), 259
Survival of the Friendliest (Hare and Wood), 11
Survivor's guilt, 158, 162
Sweden, 271
System I and System II, 99. *See also* Dual-system theories
Systemic racism, 84, 289

Tax(es)
  cheating on, 209
  fairness and, 283
  Laffer curve and, 270–271
Tax code, 204
Tax game (public goods game), 18, 35–40, 135
  bistability to, 295
  conditional cooperators in, 269
  gossip and, 67
  insurance and, 190–192
  reputation and, 65

team interactions and, 59–60
  third-party punishment and, 63
  third-party social control and, 78
Team games, 19, 20
Teams, 55–56
  complexity of, 58
  pirates operating as, 60
  social solidarity (asabiya) in, 56–60
  within teams, 58–59
  that are not genetically related, 57–58
  third-party punishment in, 61–68
Technologies
  allowing for larger cooperative groups, 81
  improvements in communication and, 280
  prescriptive science and, 10
  social structures as, 6–7 (*see also* Social structure(s))
Ten Commandments, 238–239
Thatcher, Margaret, 271–272
Thinking Fast and Slow (Kahneman), 133
Third-party mediation, 208
Third-party punishment, 5, 79, 274, 285–286. *See also* Social control
  coalition defense and, 61–62
  costs to the community, 87, 218
  of criminals, 62–63
  escalation process, 237
  gossip and, 67–68
  government and, 245–247
  human brain structure involved in, 152
  instinctual/Pavlovian circuits and, 123
  nonviolence and, 84, 85
  ostracism, 63–65
  for polluting, 189
  religion and, 230–237, 235
  reputation and, 65–67
Third-party social control, 73–74, 78–86, 149, 207–208
Thirteenth Amendment, 85, 245
Tikkun olam, 226
Tipping, in a restaurant, 50

"War on drugs," 197
Watchmen (graphic novel and movie), 61,
    308–309n6
"We all do better when we all do better," 11,
    180, 295
WEIRD (Western, educated, industrialized,
    rich, and democratic), 129, 178, 314n4
Wellstone, Paul, 11, 295
West Wing (TV series), 216
Whaling, 79–80
Whistleblowers, 82
Wilkerson, Isabel, 11
Williamson, Jack, 259
Williams, Serena, 20
Willpower, 196, 197, 198, 200
Wilson, David Sloan, 5, 38, 39, 225, 234, 239
With Folded Hands (Williamson), 259
Withholding taxes, 204
Within-group communities, 260
Within-group dynamics, 41, 130, 152, 239,
    274
Worker bees, 57
World War I, 156
World War II, 103–104, 108, 115, 156,
    163–164, 253, 275, 288
Wright, Daunte, 122

Xenophobia, 134–135
    group competition and, 276, 277
    intrinsic goals and, 130
    parochial altruism and, 109, 123, 179, 192,
        276
    purity and, 229

Yanez, Jeronimo, 121, 122, 314n20

Zero-sum games. *See also* Non-zero-sum games
    about, 19, 20
    assurance game compared with, 30
    empathy and, 145–146
    in *The Princess Bride*, 20–21